Inside Criminal Networks

For other titles published in this series, go to
www.springer.com/series/6564

STUDIES OF ORGANIZED CRIME

Volume 8

Series Editor:
Frank Bovenkerk, *University of Utrecht, Willem Pompe Institute, The Netherlands*

Carlo Morselli

Inside Criminal Networks

Springer

Carlo Morselli
School of Criminology
University of Montreal
Montreal QC, H3C 3J7
Canada
carlo.morselli@umontreal.ca

ISSN: 1571-5493
ISBN: 978-0-387-09525-7 e-ISBN: 978-0-387-09526-4
DOI 10.1007/978-0-387-09526-4

Library of Congress Control Number: 2008937031

Printed on acid-free paper

springer.com

Acknowledgments

This research started and evolved under the guidance of Pierre Tremblay—as usual, I am indebted to his tremendous support as my teacher, colleague, and friend throughout the years. I have also been fortunate enough to find myself in a stimulating research environment at the Université de Montréal's School of Criminology. During the past 7 years, several students joined me in my effort to integrate the network perspective in criminology. Partnering with them definitely accelerated the completion of this book and the elaboration of the research program extending from it. For this, I thank Katia Petit, Cynthia Giguère, Julie Roy, Maïa Leduc, Juliette Jarvis, Catherine Rochefort-Maranda, Karine Descormiers, Isa Savoie-Gargiso, and Magali Lehoux-Richer for their collaboration at various phases along the way and, particularly, for showing me that I was not alone in my enthusiasm.

The fieldwork for this book took place during separate phases that coincided with each of the case studies presented. Throughout this overall time span, I had precious assistance from a number of people working in law-enforcement and later criminal justice circles. From the Montreal Police, Claude Rochon, Yvan Delorme, Pierre Daoust, Robert Laplante, and Gérard St-Jean helped me take my initial steps by facilitating access and providing me with helpful sources for the first case studies in this research program. Ten years later, and also from the Montreal Police, Mathieu Charest, Maurizio D'Elia, Michelle Côté, and Gilbert Cordeau provided me with similar help and collaboration in the preparation of the last case study in this research. The most surprising event throughout the fieldwork was gaining access to the data sources at the base of the Hells Angels case study. This will likely lead to additional research endeavors beyond this book. What amounted to a massive data set and the most ambitious of the case studies in this book began with a simple telephone call, a straightforward question, and an equally uncomplicated reply. For this, I thank Lise Rochefort and Jacques Bouchard for their candor and for inviting me to help out in the clean up. Managing such data was also simplified by my friend, Hakim Alj, who prepared a quick computer program to transfer wiretap data into a neat network matrix.

As the chapters in this book were being completed, a number of colleagues took the time to read the ongoing work and shared their comments and advice with me, improving the book in the process. I thank Maurice Cusson, Martin Bouchard, Benoît Dupont, Edward Kleemans, Nicholas Dorn, Frank Bovenkerk, and Marcus Felson. During the last stages of this book's preparation, Klaus von Lampe united an impressive group of researchers and organized a conference on criminal networks. This seminar provided me with the ideal forum to plunge into and take stock of what I was about to complete, while also sensitizing me to the many questions and research paths that were left unexplored.

Funding for this research came from three organizations. Between 2001 and 2004, the *Fonds pour la formation de chercheurs et l'aide à la recherche* provided the necessary financial support with a junior researcher grant. Additional financial support was subsequently provided by the *Social Sciences and Humanities Research Council* between 2004 and 2008. Financial assistance was also provided by the Research and Statistics Division of Justice Canada—I thank Valerie Howe, Nathalie Quann, and Damir Kukec for their help coordinating this final segment.

Some of the sections in this book have already been published in various articles throughout recent years. I thank Elsevier Publications and the editor of the January 2007 issue of *Social Networks* for permission to include the case study in Chapter 4; Springer Publications and the editor of the May 2006 issue of *Crime, Law, and Social Change* for authorization to use the case study in Chapter 5; Taylor and Francis (Routledge) and the editor of the May 2007 issue of *Global Crime* for permission to publish the case study in Chapter 6; and Blackwell Publishing and the editor of the February 2008 issue of *Criminology* for permission to include the case study that appears in Chapter 7.

Throughout the past year, I was fortunate to find myself on sabbatical from the Université de Montréal. This opportunity gave me the time and focus needed to complete the final stages of this book. A large portion of this book was written as my family and I resided in Padova, Italy. I thank Silvio Scanagatta, Renato Stella, Patrizia Casotto, and other colleagues and staff from the Dipartimento de Sociologia at the Università degli Studi di Padova for kindly hosting my stay and providing me with the necessary academic environment and resources to work on my research. I also thank Ernesto Savona, Andrea Di Nicola, Mara Sartori, Valentina Tenti, and Francesca Belton at the Transcrime Center in Trento for hosting a final writing sprint toward the end of this project.

When I stop to think about the entire project, maybe the best part of writing this book was being able to do so in Italy over the past year. To my family and friends in Montreal, Padova, and Gorizia, thank you for the wonderful distractions. Finally, and most importantly, I thank Fabienne Cusson and our children, Luca and Béatrice, for remaining curious and sharing the many experiences along the way.

Contents

Chapter 1
The Criminal Network Perspective

The scope, forms, and contents of criminal organizational systems vary across an extended continuum. They range from simple co-offending decisions to seize an available and attractive criminal opportunity to sophisticated designs to monopolize a given market or geographical territory. They may be based on the incentives offered by a one-time partnership to execute a criminal venture or framed within a bureaucratic-like infrastructure that demands and enforces exclusivity on the actions and productivity of members. Within the range of criminal organizational systems, interactions between coparticipants can be based on family or friendship ties, background affinities, resource sharing, individual expertise, group loyalty, or governance by a dominant figure.

Which criminal phenomena remain problematic across such a continuum is difficult to tell. The lesser organizational forms may appear as trivial single events, but when aggregated they emerge as a formidable problem—note that the majority of crimes involve more than one person. The other extreme presents an immediate and serious threat—that of a vast criminal organization that governs the actions of its members who prey off society and disrupt the value systems that maintain collective order. Many would say that the latter is primarily the stuff of myth and public sensation seeking. I tend to agree, but as students of crime we are nevertheless forced to confront (or reflect on) such a possibility, regardless of how unlikely it may appear.

What is lacking in the study of criminal organizational systems is a concept that offers enough flexibility to incorporate such a wide variety. The sociology of crime has gone through its share of concepts that denote the social agglomeration of offenders from different angles. Often, the features uniting a concept were considerably influenced by the prevailing doctrine of the time. In search of the lumpen proletariat or immersed in the class-conflict framework, contemporary and historical studies of the mid-to-late nineteenth century were marked by the dangerous underclasses (Mayhew 1861/1968; Booth 1902), the anti-society (Chesney 1970), or social banditry (Hobsbawm

C. Morselli, *Inside Criminal Networks*, DOI 10.1007/978-0-387-09526-4_1,

1969, 1959). Early Chicago school research was concerned with the disorganized neighborhoods of the urban core and introduced us to the intergenerational transmission of criminal values and traditions (Shaw and McKay 1942). Extending from this position, Sutherland (1947) proposed the concept of differential social organization, which distanced us from the pathological assumption underlying the idea of a disorganized subsection of society and invoked the concept that, in certain contexts, groups organize around criminal values and activities just as other groups would converge around noncriminal activities. The latter half of the twentieth century saw Sutherland's students extend his ideas on criminal values and organization toward achievement-based subcultural theories (Cohen 1955; Cloward and Ohlin 1960) and the apparent rise of the bureaucratic-like criminal syndicate or confederation (Cressey 1969). Mainstream criminology did not follow growing claims that offenders were organized or could organize into normative subcultures or systems of governance that paralleled those of official states. Instead, the mainstream experienced a shift back toward the individual. However, in the demise of selective incapacitation attempts to target habitual and serious offenders and in the rise of routine activity theory, researchers in the 1970s and 1980s were increasingly concerned with the spatial convergence of offenders, leading to the rise of environmental criminology (Brantingham and Brantingham 1984, 1981) and the revelation of crime hot spots (Sherman, Gartin, and Buerger 1989). A close and more recent relative of this is Felson's (2003) convergence setting—the place where potential offenders go to meet, solicit, or recruit established or other potential offenders. During this same period, the escalation of the war on illicit drugs led to economic-based approaches emphasizing the criminal market perspective (Reuter and Kleiman 1986). Aside from these theories and perspectives that have been put forward to understand criminal organizational systems, specific offending groups have been referred to with a multitude of terms, as diverse as gangs (Thrasher 1927), action-sets (Boissevain 1974; Walsh 1977; Baker and Faulkner 1993), combinations (Block and Chambliss 1981), crews (Adler 1985/1993), firms (Reuter 1983; Hobbs 2001), and criminal/illegal enterprises (Smith 1980; Haller 1990).

Most recently, and in light of past and current evidence that we are living in a small world (Milgram 1967), a network society (Castells 1996), a connected age (Watts 2003), or a network-driven information economy (Benkler 2006), we have seen the rise of the social network concept as the principal organizational structure for noncriminal and criminal social relationships. The present book fits within the growing set of research endeavors that have turned to the network perspective for studying criminological issues. It is a follow-up to a previous book, *Contacts, Opportunities, and Criminal Enterprise* (Morselli 2005).

My choice of concept for both these books, the criminal network, is indeed influenced by our times, but it is not simply fashionable. In fact, it was through the study of various criminal operations and experiences that I came to understand that the network concept and the methods and overall framework that have developed around it, incorporate the theories and terms that have been proposed throughout the years in the study of criminal organizational systems. My use of the network configuration is therefore not metaphorical. Instead, my position is that whether in a marginalized subset of society, a problematic neighborhood, a deviant subculture, a criminogenic geographical space, or in the criminal market, networks are in place and it is within such organizational systems that criminal opportunities are generated and collaboration to seize such opportunities is organized and executed.

But not everyone agrees that networks are important for our understanding of crime. Throughout recent years, I have been approached and questioned by colleagues in regard to my adherence to the social network framework and its relevance for the study of crime. During a presentation at the School of Criminology, Université de Montréal (in January 2007), Tom Naylor, one of our better critical analysts in the field of organized or serious crime (Tom, of course, would be the first to reject both terms) and probably the most versatile researcher to take an interest in the flow of dirty money across the world, expressed his lack of understanding over what all the fuss was about in regard to social networks. Tom had just completed a book on the myths and collateral damages surrounding the international terrorist threat (Naylor 2006). In the book, he passionately deconstructed the premises at the base of the on-going war on terror. He demonstrated the limits behind the claim that the current terrorist threat is the workings of a sophisticated structure governed mainly by key al-Qā'idah members. He illustrated why it was unlikely that Usama bin Lāden was the international kingpin that many make him out to be. He provided the operational details for various bombing attacks to illustrate the simplicity and inexpensiveness of many of the attacks that have been attributed to the masterminding of top al-Qā'idah members. Naylor argued that a sophisticated organizational structure with a charismatic or absolute ruler was not necessary to execute most of the attacks that we have unfortunately witnessed throughout recent years.

Whereas Naylor convinced me (as he had previously in the area of organized crime) that crime control policies and threat assessments are often based on other agendas and vested interests, I would like to show him and others who maintain a similar critical stance why the social network configuration offers an alternative approach that more suitably fits the crime. A social network, for Naylor, was a limited construct because, in theory, it could include everyone. Naylor is partly right, but as we will see, there is more to it than that.

Marcus Felson is also skeptical when it comes to the social network framework and its application to criminology. Now, Felson and Naylor are very different people and researchers, but they share a common ground in that Felson has been explicit on the limits of the network framework in his work on co-offending patterns and the convergence settings that shape them (Felson 2003) and in his general portrait of the natural habitat of crime and offender foraging methods (Felson 2006). When it comes to organized crime and more general crime, Felson does not see the theoretical value of the network framework, nor does he see many concrete applications extending from it. As with Naylor, I would like to show Felson and others with a theoretical and preventive soul why the social network framework does have its place in the field of criminology.

Naylor's got a point. In its most complete form, a social network could include everyone (Fischer 1977: 33–34), but, as researchers, our specific questions and research objectives often bring us to study partitions of the whole. That a social network includes everyone is indeed a limited operational construct. However, that a specific social network, in theory, may include anyone from the "whole" network does offer its most important feature: a network is a flexible structure that could always generate new nodes and relationships. Networks are resilient and they may be composed of many features. A network can be, but does not have to be, class-based. It can be, but does not have to be, a product of urban design and economic conditions. Actors in a network can be, but do not have to be, geographically concentrated, formally organized, or united by a common value system. For operational purposes, the only defining feature that we have come to accept with a respectable level of consensus is that a social network is a finite set of actors and the relation(s) that define them (Wasserman and Faust 1994: 20). Because a network combines overlapping relationships, actors in the network could be directly or indirectly related. This definition does offer a wide set of possibilities, but, in essence, it is telling us that at any given time, we will be able to locate people, places, and things that are related to each other in various ways. The purposes, reasons, or relationships that unite actors in a network are what our boundaries of the network are based on and from what our theories extend.

The network framework, in short, is not a theoretical framework in itself. Instead, it offers us an analytical framework that congregates social interactions amongst a given set of actors. Once the network is assembled along one or more types of relationships, it becomes possible to assess various analytical themes that are at the center of the main theories of crime. The networks studied across this book are based on collective criminal ventures and, although, as the next section will review, relationships in crime have been found to have distinctive features, I stress that criminal networks could, in theory, include pretty much anyone—but, in practice, not everyone at the same time.

I. The Criminal Network Difference

An important challenge for every student of crime was set forward by Sutherland toward the end of his formulation of the differential association process. It is not sufficient to simply transpose theories and models from general social life to crime settings. Criminal phenomena require their own explanations. For example, acknowledging that offenders have similar needs and values as nonoffenders does not offer much in terms of an explanation of criminal behavior. The task of the criminologist is to illustrate how or explain why crime is different. Within the framework laid out for this book, my first task is therefore to establish why criminal networks are different from noncriminal social networks. Crime, after all, is a social phenomenon, but criminal networks and general criminal behavior do have distinctive features from noncriminal counterparts.

Curiously, early social network research never developed these nuances, although some of the first studies in the field were conducted in what could be described as offender settings. The intellectual background of social network analysis is generally rooted in either an anthropological tradition that searched for new ways of observing and understanding social order and political structures or an economic sociological tradition that developed the more formal social network methods and concepts that we are familiar with today.[1] Few researchers from either of these strands were concerned with crime or deviant groups. However, while rare, the relevance of social network analysis for criminology is not new.

Indeed, one of the first advocates of sociometry, Jacob Moreno, applied his pioneering techniques in the context of Sing Sing prison (Moreno 1932; see also Moreno 1957) and the Hudson home for girls, a reform school in which adolescent girls were placed under state supervision by New York State courts (Moreno 1934/1953; see also Jennings 1943). But while Moreno's studies may appear to have criminological roots, his contribution was primarily in the field of social psychology and, more particularly, for exercises in spontaneous-creativity and the development of later notions of sociatry, sociosis, and social gravitation. Overall, Moreno's clinical work in group psychotherapy demonstrated how treatment within group

[1] Scott (1991) provides a basic account of both schools. Freeman (2004) provides a more detailed and personal description of the field's intellectual history.

settings increased affinities between individuals, improved their sociability, and made their passage through the given facility more effective.

The correctional context was not incorporated in this understanding of the morphology and inner workings of such groups. For example, in Sing Sing's adult correctional environment, Moreno (1957) acknowledged that the community was in constant flux (e.g., the entrance, transfer, or release of inmates) and prison life was stated as a fundamental factor for understanding the balance between subjects (p. 9). Moreno's intention for studying prison groups was *to transform the promiscuous, unorganized prison system into a socialized community through a method of assignment of prisoners to social groups* (p. 7). His approach had the combined effect of organizing interactions between inmates by assigning them to specific groups, while rendering the administration of these groups more flexible by tolerating movement between groups in problematic cases. In assigning inmates to groups, Moreno considered factors such as an inmate's criminal record and prison behavior. His main concern, however, was to assess various forms of relational characteristics (e.g., leadership and neutral or antagonistic attitudes between inmates) in order to improve management within the prison. While re-arranging the prison's social setting was the main objective of that particular experiment, no links were made between the contextual factors of the prison and the sociometric findings that emerged from the therapy sessions. Most notably, no reflections were offered in regard to how inmate groups were influenced by the transient and highly controlled prison setting. The prison setting, in short, was assumed to be an ordinary social setting.

Similarly, in Moreno's study of social relationships in the Hudson home for girls, little consideration was devoted to how this particular context and sample influenced the structure of groups and the attractions, repulsions, and indifferences between group members. Jennings (1943), who studied this same sample of girls with Moreno, was slightly more sensitive to this environmental factor. She stressed that this school was a "closed community" and hence, *subjects are limited in the forming of relationships to other individuals of the population* (p. 26). She also described the study's sample as representing a *cross-section of the socially and economically under-privileged of the state's population as a whole* (ibid). While she maintained that the *institution's population does not deviate too radically from the general population* (p. 27), she also emphasized that the main reason for "commitment" in this particular center was "sexual delinquency" or, in contemporary terms, precocious sexual behavior. Moreno largely ignored these characteristics and both Moreno and Jennings overlooked environmental factors in interpreting their respective results. For example, that the "structure of rejection" at the Hudson home for girls was motivated by the lying, stealing, dirtiness, mean-spiritedness, and quarrelsome behavior of some of the girls (Moreno 1934/1953: 330; see also Chapter 8 in Jennings 1943) illustrates particular qualities of this sample in comparison to an open, more conventional group of adolescent girls. The structure of groups and subgroups therein were likely influenced by the manifestation of such behavior as well as by the intervention context of the setting—the "house-mother," for example, was described as a parent surrogate by Moreno, whereas she may have more likely filled the role of case-load manager or juvenile correctional educator in a contemporary context. That they were in a state facility and within a repressive setting had much to do with the level, form, and structure of sociability within the home.

In defense of Moreno and Jennings, an emphasis must be placed on the fact that their clinical research was designed to construct the groups that would consequently be studied. In this sense, the researchers were as much controllers as any administrative official that was in place before the onset of the research. This is a quite different situation to the study of groups already in place. The latter, of course, would have required a consideration of the controlled setting. This problem persists beyond Moreno and Jenning's research.

During the 1950s, John Gagnon conducted a survey of 67 inmates in a single prison. The prison survey was concerned with inmate organization, friendships, and frequent associations within the prison. Sociometric data from this survey were analyzed three times (MacRae 1960; Hubbell 1965; Moody 2001) and none of these studies made the prison context a key analytical element in their common aim of identifying subgroup formations within the inmate network. MacRae was partially sensitive to the contextual factor in that he did state that the *internal structure of this group was somewhat less clear-cut than one would encounter in residential communities or school populations*, but the value of this distinction was only acknowledged for its methodological utility. Hubbell used this same inmate data in suggesting an alternative method for identifying clique formations. He was also partially sensitive to the incarceration setting, but only for emphasizing that inmates were not restricted to interactions with closed groups. Hubbell raised the importance of acknowledging inmate interactions with *friends and relatives outside the institution, prison officials, and inmates who belong to other groups within the institution* (1965: 387). In his critique of MacRae's factor-analytic approach and its assumption of reciprocity in subgroup formations, Hubbell also stressed that reciprocity within a group was particular to the context in which people came to identify and converge—he illustrated this by arguing that movie star networks were more reciprocal than fan club or political support networks. Yet, no discussion was devoted to how the link between reciprocity and fellowship bonds developed within the prison context.

Years later, Moody also turned to the Gagnon prison survey to develop an algorithm for peer group identification. Moody restricted his analysis to the 39 inmates making up the largest component[2] within the prison. Using three different procedures, he consistently identified six subgroups within this set of inmates. Inmate designations within the subgroups were highly correlated across the three algorithmic procedures, but because Moody's objective was to evaluate the performance of his proposed algorithm in large and small networks, the use of the Gagnon prison survey was strictly instrumental. Once again, no attempt was made to assess how peer groups and social interactions in a prison may be different than those beyond that special setting.

Clearly, a critique of these studies must acknowledge the underlying clinical and methodological objectives that guided each research endeavor, but the question persists: to what extent does a highly monitored and therefore covert setting, such as a prison, shape the social interactions and organization of its members? Sykes's (1958) research of a maximum-security prison around the same period as Gagnon's prison survey would suggest that the prison context (or the total institution) is particular when it comes to internal order, solidarity issues, subgroup and role type identification, formal and informal leadership designations, and interactions with the outside world and prison officials. None of these distinctions extending from the highly controlled prison setting were ever incorporated in the social network studies of the prison. These early and later developments of social network analysis, although applied within typical criminological intervention settings, generated theoretical and analytical tools that were not adapted to the criminological objective—at least not within Sutherland's creed for the discipline.

But there is something different about such intervention contexts and general settings that unite actors before, during, or after crime. Criminal networks are not simply social networks operating in criminal contexts. The covert settings that surround them call for specific interactions and relational features within and beyond the network. This was the basis of Erickson's (1981) study of how risk and the need for security shape the structure of some secret societies. When secrecy is a "necessary condition" (p. 195) and risk is a fundamental factor, trust, personal vouching, and social tie strength all increase in importance. Baker and Faulkner (1993) also placed risk and concealment as the most important features in criminal networks. In their study of the internal structure of collusive operations in three segments of the electronic equipment industry (switchgear, transformers, and steam turbine

[2]A network component is a segment of the network in which all nodes or actors may reach (directly or indirectly) each other. A network with no isolated or separate group of nodes—all nodes in the network may reach each other—is a single component.

generators), they argued that the structures that such criminal networks maintained were driven primarily by concerns for concealing the illegal activity rather than for rendering it more efficient. Aside from initial secrecy, Baker and Faulkner also pointed to how the criminal network structure often protects some participants when their illegal activities are no longer secret and participants become the target of investigation by external forces. Increasing protection after detection can take a variety of forms: limited physical interaction between network participants; the minimization of communication channels; the creation of internal organizational buffers to detach participants from one another, and the decentralization of management to shelter leaders.

There are many overlaps between noncriminal and criminal networks, but criminal networks are not simply mirror images of noncriminal networks. Liddick's (1999) allusion to the "carnival mirror" (p. 428) is more fitting here in that it captures the amplification of some organizational features mainly due to the proscribed status of the activities that take place in criminal networks and the pressures that are continuously opposing them. All social networks face some level of control, but a criminal network faces more intensive and systematic control from external and internal forces. Such controls may be linked to the prohibited status of action within a criminal network or to dealings with coparticipants. From the outside, a criminal network is structured and adjusts to formal and informal social control agents—notably, the police, public services, and the community. From the inside, a criminal network is structured and adjusts to checks between participants that confront each other in conflicts that cannot be resolved by legitimate social control agents. These controls govern the general environment, but participants (as individuals or as a collective) adjust to function in such a context.

II. Flexible Order

Although criminal networks are constrained settings, it is also true that any network participant may control how others control them. How criminal networks take shape in light of the internal and external controls confronting them and how individual positioning within such networks varies are the starting points of this book. Research on organized crime and general illegal enterprise has made these key issues. Past efforts to resolve them have typically been formulated within the scope of an ongoing hierarchy-decentralization debate. Over four decades, this debate has grown to contain more than two positions. Some argue in favor of the stereotypes that have been assumed to structure high-level crime by referring to the presence of tightly knit, hierarchical, monolithic, and intensely regulated criminal organizations in a number of activities (Cressey 1969). Others have confronted this position by emphasizing the difficulties for criminal collectives to organize in such overly regulated forms and insist, instead, on the loose and decentralized nature of most criminal groups (Block and Chambliss 1981; Reuter 1983; Haller 1990; Potter 1994). In contrast to the opposing argument, this position emphasizes that offenders are simply not abiders of rules and they consequently organize themselves in ways that are the least possibly regulated. Another set of researchers have attempted to merge hierarchy and decentralization by introducing a transitional phenomenon—criminal settings that were traditionally hierarchical became more decentralized once loyalty and formal order were displaced by individual expertise and an entrepreneurial spirit (Arlacchi 1983; but see also Cressey 1969).

Finally, some recent studies of organized crime and illegal market settings have turned to the network model in suggesting an alternative framework that embraces the possibility of both the organizational confines associated with the hierarchy and the flexibility maintained within more decentralized configurations (McIllwain 1999; Raab and Milward 2003; Morselli 2005). Waring (1993) and McAndrew (1999) tell us why the network is a suitable alternative to traditional terms used to denote criminal collectives, particularly the group. First, in co-offending settings, many offenders may be part of the same criminal operation, but they never actually meet. The network framework grasps this scenario; the group does not. Second, the network framework provides what is probably the most suitable working tool for assembling the resource pooling that is typical of entrepreneurial crimes that require a greater level of skill, sophistication, and complex collaboration. Third, networks represent collectives that are more adaptable than groups. The network's adaptive features are in its flexibility, which is conducive to the hostile and uncertain environment generally associated to crime.

There are, however, different ways of applying the network concept. Von Lampe (2003) distinguishes two interpretations. One interpretation places the network configuration in the middle of an organizational continuum, with the isolated individual at the far left and the formal criminal organization at the far right. From this perspective, the network is a form of organization that is independent of other organizational forms (Best and Luckenbill 1982 offer a similar organizational typology). The second interpretation would have the network transcending all forms of organizations. Within this representation, formal organizational systems may emerge from the network. What this alternative suggests is that a network is sustained even though fixed and finite organizational forms therein may disperse over time.

This latter interpretation is consistent with Ebel, Davidsen, and Bornholdt (2003), who argue that criminal networks are complex settings that do not emerge from predetermined designs but within self-organized processes that emerge from the interactions of participants. Kleemans (2007; see also Kleemans and Van de Bunt 1999), in a description of criminal networking in the Netherlands, also presents a similar "social snowball effect" in which people come into contact with participants in an ongoing criminal operation. These new arrivals become participants themselves and, in later ventures, serve as entry vectors for subsequent people who are attracted to those criminal operations.

These ongoing interactions in criminal networks combine to create a context of flexible order. The idea of flexible order begins with the assumption that there is common ground to be found in the interaction between individual and collective interests. A second claim emphasizes the bottom-up organizational force of individual interactions and that a central governing authority is not a necessary condition for reaching social order. In brief, the network is a self-organizing structure that is essentially driven by the emergent behavior of its parts. Such parts may include a variety of actors, subgroups, clusters, and even hierarchies that are absorbed in the overriding network structure. The idea is not new. It is consistent with theories of social order ranging from Hayek's (1973) spontaneous (or emergent) order and Axelrod's (1983) cooperation equilibrium. The idea is also coherent with the most consistent assessment of conventional networks: the social network's force is the flexibility that it offers its members. Networks have been described as "lighter on their feet" (Powell 1990), in contrast to slow moving and fiat-filled hierarchies.

The idea of flexible order is also compatible with the opportunistic, short-term, and impulsive features generally associated with offenders by mainstream criminologists. However, there is a key difference between the assumptions underlying flexible order and those on which the principal criminological theories are based on. What this thesis tells us is that it is not so much that individual offenders lack the skills or competency to plan and structure their actions; instead, it is more that the organizational systems in which many find themselves when executing their crimes and collaborating with other offenders do not require such extensive planning and long-term organization. What participation in crime requires is a capacity to react quickly and networks are the organizational systems in which such reactions are most suitably played out. Positioning and remaining flexible is the key. From outside the network, such flexibility may appear to be mere opportunism, short-term thinking, or the result of uncontrolled impulses. From inside the criminal network, however, quick reactions and adjustments are precisely what are called for.

III. Centrality and Key Player Designations

Stating that a network is built on flexibility and lacks formal organization does not mean that the network cannot be centralized around certain poles of action. The extent to which a network is centralized around one or a small set of actors has consistently been the first step in most social network analyses. From its initial conceptualization, centrality was associated with some form of influence or control (Bavelas 1948). Freeman (1979) has been at the center of such research and his contribution has been most important in clarifying and operationalizing the concept in its various forms. Centrality, at the individual level, could be measured simply as the number or proportion of contacts with whom a participant is directly connected within a network. Referred to as degree centrality, this measure reflects the extent to which a network participant remains in the "thick of things" (Freeman 1979: 219). Another common measure is that of betweenness centrality, which adjusts one of the main limits in the degree centrality measure—that some participants may have a lower degree of direct contacts in the network, but they are nevertheless centrally positioned as key intermediaries along the shortest paths that unite network participants. This is particularly important in less cohesive networks. Participants with high betweenness centrality in such networks have consistently been referred to as having a strategic position (see Freeman 1977 for a review of this literature). In contrast to those with high degree centrality, these participants bring a brokerage dimension to the network and they are argued to have an advantage when it comes to controlling information or communication flow (Freeman 1977, 1979).

Direct Centrality and Visibility

Although the premise that centrality is an indication of importance, influence, or control in a network may appear valid, it is also contestable, particularly in criminal contexts. Whether centrality is an indication of power has been the subject of much discussion, theoretical reformulations, and methodological advancements in the social network field. In the field of criminal network analysis, several researchers have also reflected on the implications of carrying over centrality measures in the study of crime.

What does it mean to be central in a criminal network? For Sparrow (1991), the central node held a position of strength within a network. Thus, removing that strength would weaken the network. Peterson (1994), in contrast, maintained that nodes positioned at the center of the network were the most connected participants and therefore the most vulnerable (p. 31). Centrality was therefore an indication of weakness. Williams (2001) challenges the central node issue even further by questioning whether most criminal networks have a sensitive center of gravity to begin with (p. 71).

One study which provides empirical evidence to address whether actor centrality is a sign of strength, vulnerability, or nothing at all is Baker and Faulkner's (1993) analysis of price-fixing conspiracies. Within the three collusive networks, they found that the participants who had higher degree centrality were more likely to be found guilty and receive greater sentences, leading the authors to conclude that being in the thick of such conspiracies made a person vulnerable. What Baker and Faulkner's findings suggest is that central positioning in a criminal network is primarily an indication of a participant's visibility and the more strategic participants in the network would therefore be more content in remaining peripheral, and not central, to the action taking place.

Within this same study, Baker and Faulkner found that the more decentralized networks were more likely to yield a higher proportion of guilty verdicts among the targeted participants. The authors present this as a paradox in that they initially expected decentralized networks to be more protective of its members. They resolved their puzzle by assessing the core–periphery structure of centralized and decentralized networks. A centralized network is composed of a small, densely knit core and a large, dispersed periphery. In such a structure, participants at the core are particularly vulnerable because of their high degree of connectivity. Participants in the periphery are better insulated because they maintain few contacts with each other and with core participants. Thus, most participants in the periphery of a centralized network avoided a guilty verdict because they remained distant from the collusive actions—in other words, they were less directly implicated. In a decentralized network, a visible core is lacking, and direct involvement is more evenly distributed across the network. Thus, if a decentralized criminal network is detected and intervention takes place, guilty verdicts are likely to be distributed more evenly across the length of the network.

Baker and Faulkner's explanation for their findings is consistent with more recent research on network resilience and vulnerability. Watts (2003), for example, illustrates how connectivity in network systems makes them vulnerable to cascade failures while at the same time offering the structural resources that allow them to resist and rebound. In contrasting scale-free and random networks, Watts recounts the work of Albert, Jeong, and Barabási (2000) and summons the irony uniting the strengths and weaknesses of network structures in the context of random or deliberate attacks against scale-free or random networks. Scale-free networks have a central core. They are resistant to random attacks because the majority of less central nodes will more likely be affected and the loss of such peripheral nodes for networks with central hubs is less significant for the survival of the network. In contrast, random networks are more decentralized and less resistant to random attacks because the loss of any single node will be more important to the remainder of the network. But Baker and Faulkner's networks were not subjected to random attacks—they were targeted deliberately. In the context of deliberate attacks, network vulnerability is inversed: central nodes are more likely targeted, making scale-free networks more susceptible than random networks (Watts 2003: 191).

What such research suggests is that the controls that were set upon the networks in Baker and Faulkner's study were typical of most attempts to contain criminal networks. Because targeting generally converges on the more obvious, visible, or central participants, such controls will fall short when confronting a decentralized network and will miss out on important peripheral participants when confronting a centralized network. The bottom line is that knowledge of a network's structural features must be obtained before any form of intervention is applied.

Further research on criminal network disruption has raised other aspects that must be addressed when assessing centrality in such structures. The first concerns the presence of other important participants in a network who are not directly central. Carley, Lee, and Krackhardt (2001) provided an additional nuance within the context of disruption strategies by illustrating how the centrality approach may be misleading, particularly in decentralized networks. They distinguished between central participants and leaders. A leader was defined as the participant with the "highest cognitive load" or the network member who manifested the most qualities associated with leadership potential (e.g., prior experience, cognitive ability, extroversion, resourcefulness, high stress tolerance, strong self-esteem, openness to new experiences, and willingness to delegate tasks). The central participant was defined as the node with the most contacts (or highest degree centrality) within the network. In a network in which centrality and leadership are occupied by two distinct participants, extracting the central node would not necessarily lead to the expected disruption of the network and the extraction of the leader would not necessarily result in the emergence of the central node as the new leader.

A second aspect concerns adaptation within the network. Participants in network structures adapt whether key nodes are removed or not and Carley et al. stressed that an understanding of disruption effects and response strategies within the network required an appreciation of such dynamics. Using a simulation approach, they illustrated how disruption effects across a network may result in the decentralization of a once hierarchical network, a competition for leadership between remaining nodes, or the demotion of leaders.

For both central node issues and vulnerability/disruption assessment, Carley et al.'s study is valuable in that it suggests widening the scope of disruption effects when networks are manipulated by external forces. This is crucial for criminal network analysis in that we know that a basic impact that law-enforcement controls have on criminal market settings is the restriction they impose on the emergence of sizeable, enduring, and tightly knit organizations. Criminal groups operating in such settings are more likely small, short-term, and loosely organized because the consequences of illegality block ambitions for the creation of enduring corporate-like organizations (Reuter 1983). Thus, void of impunity forces, criminal markets are more likely to be competitive than monopolistic and criminal networks that emerge from such markets are more likely decentralized than centralized.

What such insights and research tell us is that the removal of a central participant or disruption at the core does not necessarily lead to increased vulnerability within a targeted network. After years of random buy-and-bust interventions, law-enforcement controls of serious crime networks have gradually come to follow the key player strategy. In drug-trafficking contexts, for example, this strategy is often rooted in the "Mr. Big" assumption that has been subjected to repeated criticism in empirical research in this field (Reuter and Haaga 1989; Dorn, Murji, and South 1992; Eck and Gersh 2000; Pearson and Hobbs 2001; Zaitch 2002). A more accurate appraisal of the social organization of drug-trafficking and other market crimes would follow a resource-sharing model in which collaboration amongst resourceful individuals would be at the base of coordination in such operations. Such an appraisal emphasizes the transactional basis of these criminal settings and is consistent with Cook, Emerson, Gillmore, and Yamagishi's (1983) rendition of a social network as the backdrop for exchange in which each actor has resources that other actors in the network value, each connection between actors is an opportunity to share and profit from such resources, and the ensemble of connections in the network represent the opportunity structure for all taking part in the exchange process (p. 279).

Such a process could be based on loose or tight coordination, but what is important to retain is that all participants share a stake in the outcomes of exchange. This transactional, resource-sharing model differs considerably from the Mr. Big scenario, in which coordination and the general cause/outcome are centralized in the rigid governance of one or a select few. In drug-trafficking and general criminal enterprise research, remaining flexible to the needs and interests of all participants positioned within the opportunity structure emerge as more important.

Brokers as Key Players

Whereas visibility is a property that is best captured by degree centrality, betweenness centrality brings another sense to the network, one of strategic action that is better suited for the risk/efficiency trade-off that is generally contemplated when organizing criminal ventures. Betweenness centrality fleshes out the intermediaries or the brokers within a network. This and other measures of brokerage are not simply alternatives for the direct connectivity accounted for by degree centrality. There is growing evidence in research on legitimate organizations that key players do not maintain authoritative roles, but brokerage positions that bring flexibility, integration, and creativity to the ensemble of an organization and that benefits the individuals occupying such positions (for a complete review of such research, see Marsden 1982; Burt 1992, 2005: Chapter 1; see also Cross and Parker 2004).

A broker is positioned between disconnected others within a network. These disconnected others may occupy different hierarchical roles within an organization or they may be members of different organizations that come together for a given operation. The most straightforward proposition regarding social interaction in either context is that *brokers do better* (Burt 2005: 7, 11) because others come to rely on them.

Why do brokers do better and why is brokerage advantageous at the group level? Burt (1992, 2005) explains that people in such a position maintain a competitive edge because they control the information asymmetries that make up entrepreneurial networks. Participants on either side of the brokerage position rely on the broker for indirect access to resources beyond their reach. The broker is pivotal within such a social configuration and profits from the reliance of others. In turn, the group that emerges around the broker benefits overall because the broker extends the collective venture to wider reaches and a greater variety of opportunities. Thus, brokerage positioning is capital for both the collective and participants occupying such positions.

Within the context of criminal networks, brokers or participants high on betweenness centrality practice what appears to be the suitable trade-off between efficiency and security. There is not much evidence to support this claim for the moment, but we do know that in Baker and Faulkner's (1993) study, betweenness centrality had no significant effect on either verdict or sentence length. This, of course, does not tell us that participants positioned as brokers were less likely to be found guilty or sentenced for longer periods. But there is some evidence from research on a variety of criminal settings that brokerage is beneficial in criminal enterprise as much as it has proven to be in legitimate enterprise.

Whereas the criminal broker does not fit the stereotypical image of the forcible and violently prone ruler of a criminal organization, his/her position in a given network does prove valuable for resource pooling and coordination. Coles (2001), Klerks (2001), and Williams (1998) reviewed past research on brokerage in crime and concluded that the presence of multiple brokers in a criminal network is more likely in groups that indicate a higher degree of sophistication or organization. The value of brokers has also been a consistent finding in studies of illegal drug-trafficking (Desroches 2005; Natarajan 2006; Pearson and Hobbs 2001; Zaitch 2002), human smuggling (Zhang 2008; Kleemans and van de Bunt 2003; Zhang and Chin 2002), stolen-vehicle exportation (Bruinsma and Bernasco 2004), and general criminal enterprise settings (Finckenauer and Waring 1998; Haller 1990; Morselli 2005).

Boissevain (1974) elaborated considerably on brokerage as a career model. His observations are relevant for the criminal career and the presence of participants occupying such a position in criminal networks. Becoming a broker begins with a *desire to manipulate people as resources in order to move ahead*; it requires *a high measure of centrality* from the start and *a certain amount of power and prestige and the time to service his relations*; the main challenge for the potential broker is *how to get people to make use of him* (p. 163). Brokerage, in this career model, is a building process which places an individual in interaction with other brokers and other players of higher and lower prestige. This interaction process continues until a particular form of network is gradually cultivated. What Boissevain suggests is that brokerage is an off-shoot of high direct connectivity, but as an actor evolves in his positioning amongst others, s/he gradually learns or adjusts to the benefits of playing the intermediary. Brokers are neither patrons nor clients. They play in between and what past research has demonstrated is that individuals who are capable of maintaining such a stance are generally well-respected, higher achievers, and strategic participants in the networks that surround them.

At the same time, network participants are not necessarily aware of the network that they are a part of; nor are they necessarily aware of their position amongst others. As Boissevain clarified, people in brokerage positions are not always conscious of their advantage and the influence that they may have on others: *in discussing brokers we must accept as a given quality a willingness to manipulate other persons, although in some cases—and here I am thinking of certain academic colleagues—the brokers concerned are not fully aware of the degree to which they in fact manipulate others* (p. 154). Boissevain's study, however, focused on people who were providing brokerage services and who profited from such a *role*. But there is more to brokerage than such roles. Much of the ambiguity that surrounds a broker's self-awareness of her/his edge has to do with the fact that being a broker is not simply a role that can easily be identified, as would be the case in occupations such as a stock broker, a real-estate broker, or even a power broker—in such cases, the occupation or role defines the *position*. Many people who do not have a broker role are nevertheless positioned as brokers in their own network settings (e.g., Boissevain's academic colleagues). Brokerage, in this sense, is above all a position amongst others and to have a full awareness of this position requires a full grasp of the ensemble of interactions around a person. Such a grasp is not easily visualized by the common eye and this is where social network analysis becomes a necessary framework.

IV. Seek, Rather than Assume, Structure

My objective in this book is not to arrive at a tightly formulated theory. Instead, my general aim is to lay down a working framework that extends from past research and proposes a series of methodological steps, analytical paths, and findings that could subsequently be followed and improved through more systematic research endeavors. The previous sections in this chapter have laid out the conceptual framework with which I approach the various case studies making up the remainder of this book. If flexible order is the common theme guiding these case studies, centrality measures are the recurring operational constructs applied across these chapters. Previous research suggests that whereas direct connectivity (or degree centrality) has gradually come to signify visibility and vulnerability within a criminal network, brokerage positioning (or betweenness centrality) has emerged as a possible sign of strategic involvement in criminal networks.

But what happens when direct and indirect connectivity are centralized around the same people? In many networks, degree and betweenness centrality are highly correlated and the distinction between the two is largely meaningless. In such contexts, it would be safest to follow previous suggestions that security or insulation is the most important dimension of criminal network participation. In this sense, when a participant is both high on brokerage capital and direct connectivity within a criminal network, s/he is above all visible. The risks associated with degree centrality, in this sense, outweigh the brokerage edge.

In this book, I have tried to avoid the pitfalls commonly met when assessing organized crime and more general group-level crime. It has generally been contested that central actors design a criminal organizational system that will allow them to control a vulnerable or disorganized mass of individuals. Flexible order does not dismiss the possibility that some participants in a given structure may emerge as more central. Instead, the proposition is that if one or more central actors emerge in a given social setting, it is as a product of the dynamic processes that take place within the structure of interactions between individual actors that are not necessarily vulnerable or disorganized to begin with. Such networks, once again, cannot be planned in advance—they are emergent.

It may be argued that, in the context of criminal activities and with the avoidance of formal borders and regulations being amongst the principal organizing forces for criminal coparticipation in local and across international boundaries, a criminal network phenomenon has always been in place in some form or other. But this is more likely an emergent phenomenon and not an intentional design. The latter, unfortunately, has been the dominant assumption guiding research on globalization and its impact on crime. Castells (1998/2000), most notably, argued that a network *system* had been put into place by diverse criminal organizations from across the world with the principal objective being greater efficiency and prosperity in an increasingly globalized and technology-driven world. A flexible order model does not require the creation of such a system. In this sense, Castells may have been correct in establishing that crime was the likely precursor to the global network trend that emerged during the end of the twentieth century (p. 210), but his belief that a global criminal economy driven by a global agglomeration of established criminal corporations from countries as diverse as Italy, the United States, Colombia, China, Japan, Jamaica, Turkey, and Russia was misguided by a small set of studies, journalistic reports, and United Nations commissions that ignored the more focused, yet less sweeping and eye-catching, research on criminal markets, organization in crime, and criminal enterprise.

The problem with Castells' assessment of the global "threat" and those studies which he drew from is that the structure of crime was never sought—it was assumed. Castells began by defining "global crime" as *the networking of powerful criminal organizations, and their associates, in shared activities throughout the planet* (p. 170). He later stated that this new phenomenon was well documented amongst the sources that he followed, but *largely ignored by social scientists, when it comes to understanding economies and societies, with the arguments that the data are not truly reliable, and that sensationalism taints interpretation* (p. 171). Castells stated what many others have argued in the past: *I take exception to these views. If a phenomenon is acknowledged as a fundamental dimension of our societies, indeed of the new, globalized system, we must use whatever evidence is available to explore the connection between these criminal activities and societies and economies at large* (p. 171). True, the empirical pursuit is most important, but such ventures are best advanced with a careful and piecemeal approach. This generally means that the impressive theoretical sweep is lost within the more modest search for specific patterns, facts, and properties.

The threat emphasis does not cease with organized crime. Since late 2001, social network analyses of criminal contexts have been devoted considerably to examining terrorist operations and, most notably, to the one leading to the hijacking attack on the Eastern United States. With this event fresh in the minds of many, control and the dismantling of criminal networks have guided data-mining operations of law-enforcement, criminal intelligence, and other mass basins of information that may reveal suspicious patterns, possible vulnerabilities, and key players (Xu, Marshall, Kaza, and Chen 2004).

As in the case in organized crime research, some have turned to the hierarchical representation as an obvious theoretical model for terrorist activities. Farley (2003), for example, argued that network representations of terrorist operations are inadequate precisely because they do not consider hierarchy. He maintained that analyses should not only focus on dividing a network into separate components but also on cutting off key players and leaders from followers. His incentive for such an approach was that the identification of leaders would render surveillance and control more selective and cost-efficient. Farley's argument and general assumption vividly reflects the problem with strict control approaches—once again, they assume order without first assessing structure. His assumption ignores too much research to be taken as a given. This is best illustrated in his reflection on Klerks' (2001) discussion on this matter. While Farley acknowledges Klerks' warning not to assume roles and order within a criminal structure before assessing the structure, he decides to overlook Klerks' advice not because he sees any fault in it, but because he assumes that the terrorist context is unique from "ordinary criminal networks": *it seems as if terrorist networks are in fact organized hierarchically, sometimes even along military lines* (Farley 2003: 404). This claim meets an important limit within the terrorism context in that in Krebs' (2002) analysis of the hijacker operation behind the September 2001 attack, it is the *dense under-layer of prior trusted relationships* (p. 50) that is found to be at the base of the network's stealth and resilience and not the commanding control of a single or select few leader(s). This point is consistent with Erickson's general appraisal that secret societies are based primarily on prior networks. This point also underlies this book's thesis in that, in schemes of flexible order, continuity in crime at a collective level is made possible not because a criminal mastermind or a predominant criminal cartel governs the actions of those falling under its shadow, but because remnants of previous organizational forms within the criminal network are always accessible for later organizational forms to emerge.

There is much truth to Felson's (2006) observations that *much of science has to do not with the facts themselves but how they are arranged to tell a story* (fn. 9, p. 320). The case studies in this book are designed to test the claims that are commonly associated with drug distribution, stolen-vehicle operations, and formal organizational or gang settings. Law-enforcement cases, which are the main sources used throughout the analyses, are excellent for doing so since they are typically inclined toward judgments and claims that weigh heavily against the criminal-network participants that are arrested and tried on the basis of such data and the interpretations that police investigators suggest. In all of the case studies analyzed throughout this book, there was always a claim that one participant or one small group of participants was dominating the operation—often with a firm hand and in a tightly knit organization. These are not simply claims that are systematically found across police investigations; these are also the claims on which the stereotypes, common knowledge and, unfortunately, much crime control policies are based on. Thus, testing the assumptions that underlie the interpretations of police investigators in their own cases also allows me to test the more general assumptions that generally underlie our own visions of criminal networks.

V. Organization of the Book

The next chapter presents the data sources and methods used throughout the rest of the book. The case studies begin in Chapter 3 with an illustration of the partnership model that has been the basis of much research on organized crime and illegal enterprise. Chapter 4 focuses on the efficiency–security trade-off and how this differs between terrorist and drug-importation networks. Chapter 5 studies facilitators in criminal networks by shifting the focus away from drug-traffickers and toward those participants who were active in the network because of their legal occupational status and resources. Chapter 6 offers an analysis of the dynamics and changes that take place in a drug-importation network that was heavily pressured by law-enforcement targeting. Chapter 7 centers specifically on participants who occupied brokerage positions in stolen-vehicle exportation rings and how they contributed to structuring and coordinating crime scripts. In Chapter 8, the brokerage component and the book's main thesis are carried into a more formal organizational setting with an analysis of drug-distribution activities extending from the Hells Angels organization in Quebec. Chapter 9 pursues a similar analysis in Montreal's street gang landscape. Each case study is designed to address a different theme and present diverse approaches for examining problematic issues within criminal networks. These issues will unfold as each case study is presented in subsequent chapters. I conclude with an appraisal of the criminal network perspective, the implications of this book's main findings, and an outline for future research. The book can be read as a whole or in parts, as each chapter and case study is also meant to stand alone.

Chapter 2
Case Study Sources and Designs

All but one of the criminal networks studied in this book operated in and around Montreal, Quebec. The exception is Valdis Krebs' case study of the hijacking network that attacked the Eastern United States on September 11, 2001. Aside from Krebs' (2001) study, which is used for comparative purposes, all other networks are the product of a data gathering quest that spanned the past ten years as I scrambled my way across police and court circles and seized every opportunity I could to access a variety of investigative cases that offered the makings of a criminal network analysis. Through access offered by investigators, police analysts, border service administrators, prosecutors, and defense attorneys, I was able to gather extensive information on ten police investigations and court proceedings. These investigations combine to form the empirical foundation for five strategic cases with which various issues are addressed in the next seven chapters. Aside from one case that deals with stolen-vehicle exportation, all others are based on illegal drug distribution at various levels.

This chapter begins with a brief description of how the various data sources used throughout the case studies were accessed. This is followed by a description of each case, a presentation of the overall research design, an elaboration of the main network measures applied in later analyses, and a discussion of the challenges of using law-enforcement data for network analysis. While most material dealing with the case studies and designs are presented in this chapter, additional details will also be included during specific analyses in later chapters.

I. Case Selection and Access

Scrambling to obtain data is indeed the most honest assessment that I can make of my experience. There is no well-planned research strategy that justifies why the cases studied in this book were selected. These decisions were

C. Morselli, *Inside Criminal Networks*, DOI 10.1007/978-0-387-09526-4_2,

largely beyond my control. It was more a question of what data I was able to access in law-enforcement and wider criminal justice circles. There is also no systematic strategy to gain access to such data. The first case I obtained was Project Caviar, an investigation that targeted a hashish and cocaine importation network operating from Montreal (see below for more details). This occurred in 1998 when I was authorized by the Montreal Police's second-in-command to enter their Narcotics Squad offices for fieldwork purposes. At the time, I was looking for a fieldwork setting for my doctoral project, and my intention was to gather enough data to create a general portrait of the cocaine market in the city. This project never came to be, largely because I was side-tracked by the possibilities of pursuing the network research and because most of the investigators working in these offices were wary of my presence.

In the end, gaining access to Project Caviar was the event that terminated any chance of conducting the cocaine market study. After about five weeks consulting investigative files and interviewing those few investigators who were willing to talk to me, the investigator who had been asked (forced) to supervise me was growing weary of taking care of me. At one point, he asked me if I was interested in seeing a few boxes of electronic surveillance transcripts and case descriptions from an investigation of a cocaine and hashish importation network that was recently completed and in which he was personally involved as an investigator. These boxes contained the evidence submitted during the trials of participants in the drug-importation network. I began examining the contents of the boxes and after assessing the logs of the electronic surveillance data, the initial design for a criminal network research study came to be.

The amount of information that had to be coded from Project Caviar was extensive and because my presence in the Narcotics Squad offices was making most of the investigators uncomfortable, I asked my supervisor in the squad if I could take the data to the university to work on. While the Canadian Criminal Code (Article 193) does stipulate restrictions that block access to general surveillance data, there are no legal constraints that restrict access to the data that I was offered. If the transcripts of the electronic and physical surveillance data were submitted as evidence during the trials of the criminal-network participants (which, in this case, some were), any citizen could simply go to Montreal's Courthouse, provide the file number of the trial, and access the data. However, because such cases are usually comprised of several boxes and thousands of pages, coding and thoroughly studying the data cannot be done in the courthouse's archive room.

My supervisor in the squad was aware that I was allowed to access such data, but he did not want me to leave with the squad's only copy of the case. Instead, he offered me his personal copy that was stored in the basement of his house, along with the copies of all other extensive investigations that he had worked on in the past. This is when I learned that many law-enforcement investigators are actually packrats in that they keep copies of the case material from their investigations—trophies of successful cases, if you will. Fortunately, this investigator had limited space in his basement and was ready to part with his documents in exchange for my confirmation that my fieldwork presence in the Narcotics Squad (and his supervision) was over. He was so happy that, in the end, he also threw in a copy of a smaller investigation of liquid hashish importers (Project *Ciel*—or Sky, in English) that he worked on soon after Project Caviar.

So, there I was, loading my tiny Renault-5 (I was a student with no children at the time) with the boxes of evidence from Projects Caviar and Ciel and ready to drive off to the university with something that could be the empirical basis for a doctoral thesis. This never became my doctoral thesis and the curious turnabout that my fieldwork took on that day would extend several more years before these and other investigative cases became the case studies presented in this book. The analysis of Project Ciel is presented in Chapter 3. Project Caviar is analyzed from different angles in Chapters 4, 5, and 6.

After Caviar and Ciel, I remained alert for other investigative cases that I could access and study. I also obtained funding from Quebec's *Fonds pour la formation de chercheurs et l'aide à la recherche* to expand my research to other cases. To avoid legal obstacles when accessing such cases, you generally have to wait for the end of the judicial process extending from the police investigation. Habitual reading of daily newspapers is sufficient to expose the different opportunities that are available. Several potential cases involving other drug-trafficking networks, illegal gambling operations, prostitution rings, or money-laundering schemes were exposed in the media, but I was never able to gain the same convenient access to the ensemble of evidence. Then I met Julie Roy, a graduate student who was interested in applying network analysis to stolen-vehicle rings that were operating in southern Quebec. Julie already had some experience in this area. To complete her undergraduate degree, she was required to do some research in a public agency (a *stage*). Her stage was with Canada's Border Services and it was here that she conducted research on the intended destinations of an ensemble of stolen vehicles seized during recent years. At the end of her stage, Julie was also given a job within the Border Services. Her position within this agency facilitated access to additional data. Together, we planned an extension of this research within a network framework. At first, all we had was information on the stolen cars and destinations. Over the following year, and with the help of police investigators who worked on the cases that led to the seizures, information was gathered to track the network participants who were operating within the province to export these vehicles. These investigative cases were referred to as Project Siren and Project Togo. Analyses of these cases are presented in Chapter 7.

And then something else happened. In late March 2001, a tandem task force uniting investigators from municipal, provincial, and federal Canadian jurisdictions began a crackdown on the Quebec Hells Angels with almost 150 arrests. This event and the investigation leading up to it have become known as Operation Springtime 2001. It was huge in every sense. Not only were these arrests the product of a massive police investigation that had been progressing for over 4 years, but it would also lead to the first application of Canada's recently enacted *gangsterism* legislation. Operation Springtime 2001 was the result of the most extensive police investigation and would lead to the largest criminal court case in Canadian history.

Once again, patience is required as one has to wait for the end of the trial process in order to access data from such cases. In the Operation Springtime case, this process took 3 years after the arrests were made. Most of the individuals accused in the case pleaded guilty and when the trials were over I applied for and received more funding to extend the case study analyses (this second funding organization was the *Social Sciences and Humanities Research Council*). I began exploring ways of accessing the massive data set submitted during these trials. Going to the courthouse was not an option since the Operation Springtime case made Project Caviar look like a simple buy-and-bust. Access had to be obtained to the complete set of computer files that contained the evidence compiled against each of the accused. Although officials in Quebec's provincial police, the Sûreté du Québec, had initially accepted my request to access this material, when the time came I was unable to procure the data. Prosecutors involved in the trial of the Hells Angels and their associates were also unwilling to share their copy of the evidence. I even asked the judges, who also denied me access.

The door to the data was finally opened for me by a couple of defense attorneys from the Hells Angels trials. Following a tip from Daniel Sanger, a freelance journalist who was researching his own book on the main informant during the investigations that led to the March 2001 arrests (see Sanger 2005), I began contacting one defense attorney at a time. By my fourth try, the strategy paid off and the timing could not have been better. The trials of the Hells Angels members and associates had gone on for a prolonged period and now that it was all over, the attorneys were told to remove their documents from the courthouse in order to make room for the next trial. I was told that if I came to the courthouse the day after, I could help with the cleaning and take whatever I found to be of value for my research. So, the next day, I drove to the courthouse, backed up my minivan (I was now a professor with children), and proceeded to load everything I could lay my hands on.

Although the contents of these documents have proven helpful as a starting point for other research endeavors beyond the scope of network research, the most important data source that I received that day was a portable hard disk that stored all the evidence compiled and submitted to defense attorneys for the trials of all individuals that were arrested and accused roughly three years previously. This hard disk included the computer-based files of the surveillance logs that permitted a more efficient processing of that data. The network surrounding the Hells Angels and their coparticipants who were targeted during Operation Springtime is examined in Chapter 8.

The last investigative cases that I examined proved to be less complicated in terms of access. Once I began completing some of the first case studies (Ciel, Caviar, Siren, and Togo), I was asked to present my research designs and results in various police settings. With many of the Hells Angels now in prison as a result of Operation Springtime, police concerns (particularly in Montreal) shifted toward street gangs. After presenting the case studies on the drug-importation and stolen-vehicle exportation networks to a group of researchers working for the Montreal Police, I was invited to apply the various techniques in a study of the street gang landscape in Montreal. I was assured access to three recently completed investigations for which judicial procedures were not yet terminated (the names of these cases therefore remain confidential). The data from these investigations included the electronic and physical surveillance data that I had become accustomed to working with. Aside from such data, I was also allowed access to the central criminal intelligence base in order to explore other sources for creating links within the criminal network settings. Curiously, I found myself in the same place as 10-years previous, when I was trying to access data within the confines of the Montreal Police's Narcotics Squad. This time, however, the door was left wide open. Along with two analysts working for the Montreal Police and with access to the surveillance records and the central intelligence base, we reconstructed the network that was targeted during these three investigative cases. This case study is presented in Chapter 9.

II. Case Study Descriptions

Each criminal network that serves as an empirical back drop to the analyses in this book is a strategic case (Baker and Faulkner 2003, 2004). Such cases allow us to assess claims regarding the patterns and inner-workings of criminal networks. They are, in short, occasions to test past propositions on the structure of crime and opportunities to explore new analytical paths. Such specific cases, at the same time, offer the potential to contribute to more general theories and perspectives on organized crime, co-offending, and the broader sociology of crime.

Because the reconstruction of these networks was based primarily on physical and electronic surveillance data, they may be best described as intercepted communication networks. These networks vary not simply in terms of the criminal activities that were involved, but also in terms of size and content. For example, whereas the majority of the networks represent operations to execute specific tasks (e.g., drug-importation, stolen-vehicle exportation), one exception (Operation Springtime) represents the wider set of criminal activities surrounding a specific organization (the Hells Angels). While the network in this latter case was firmly entrenched within illegal drug-distribution activities, the investigation that targeted it tapped into an ongoing system surrounding this organization. This system included not simply drug-importation and domestic distribution channels, but also conduits revealing the management and movement of profits from the illegal drug trade. As the preceding section and the following case descriptions illustrate, Operation Springtime is a special case in itself and will eventually be the subject of more analyses than those presented in this book.

With the exception of Kreb's representation of the 9/11 network, all names of participants revealed in the various networks remain confidential. Whereas the names of all participants in Krebs' terrorist network have been heavily publicized since 2001, only a small set of participants' names from the Ciel, Caviar, Siren, and Togo networks were publicized, and such publicity was largely restricted to the Montreal media during brief periods in the 1990s. For the gang-based network, the confidentiality of participants' names is consistent with the confidentiality of the overall investigative case. For the Operation Springtime network, the names of most members of the Hells Angels and their more reputed associates were publicized heavily during the 3-year period leading up to the end of the trials. Although this case attracted much more attention than did any of the other Montreal-based cases, the notoriety of the participants in the Hells Angels case remains largely localized to the Quebec context and I saw little value in contributing any further to their already blown-out-of-proportion exposure. In the end, I felt that there was no reason to name names when the main objectives of the research were to reveal patterns and arrive at a general understanding of how these networks were organized.

Krebs' Terrorist Network

Krebs' (2001) case study of the 9/11 events is used for comparative purposes in this book. Indeed, his representation of this network has been used widely throughout past research on covert and dark networks. However, a network built for a suicide mission is fundamentally distinct from the other "profit-oriented" networks assembled for this book (see Chapter 4). To reconstruct this network, Krebs (2001) used the growing information published in major newspapers and made available through internet sources during the weeks following the attack. The 9/11 network was composed of 37 participants, including 19 hijackers who executed the attacks and an additional 18 contributing coconspirators who did not enter the airplanes on that day, but who played key roles in planning and transmitting information during the months (or years) leading up to the attack. Krebs referred to the hijackers as the network's action segment, and to the coconspirators as complimentary participants in the network.[1]

Project Ciel

Project Ciel is based on a small drug-importation network that was importing liquid hashish from Jamaica to Montreal. This network was targeted by the Royal Canadian Mounted Police and the Montreal Police from May 1996 to June 1997. Typical of many Canadian investigations of drug smuggling and trafficking, the operations in Project Ciel were described as taking place within a tightly governed organizational framework—a hierarchy, in short. Reports from the investigation maintained that the main target of the investigation was the "organizational leader." Other key targets included the leader's "lieutenant" and a series of other subordinates. The investigation produced three separate seizures, with two taking place at Mirabel airport near Montreal and another occurring at Sangster airport in Jamaica. Overall, 75 people fell into the surveillance net. A selection process that was aimed at identifying only those individuals who were active in the drug-importation operations resulted in a final network of 25 participants.

Project Caviar

Project Caviar was a unique investigation that targeted a network of hashish and cocaine importers operating out of Montreal. The network was targeted between 1994 and 1996 by a tandem investigation uniting the Montreal Police, the Royal Canadian Mounted Police, and other national and

---

[1] A slightly modified version of Krebs' (2001) study was published about a year later (see Krebs 2002). Though Krebs was more advanced in his data gathering (this later article included 26 more participants in the overall terrorist network), no details were offered to substantiate these data modifications. Our analysis therefore relies on the first article (Krebs 2001).

regional law-enforcement agencies from various countries (i.e., England, Spain, Italy, Brazil, Paraguay, and Colombia). The case is unique because it involved a specific investigative approach that will be referred to as a "seize and wait" strategy. Unlike most law-enforcement strategies, the mandate set forward in the Project Caviar case was to seize identified drug consignments, but not to arrest any of the identified participants. This took place over a 2-year period. Thus, although 11 importation consignments were seized at different moments throughout this period, arrests only took place at the end of the investigation. What this case offers is a rare opportunity to study the evolution of a criminal network phenomenon as it was being disrupted by law-enforcement agents. The inherent investigative strategy permits an assessment of change in the network structure and an inside look into how network participants react and adapt to the growing constraints set upon them.

The principal data source was comprised of information submitted as evidence during the trials of 22 participants in the Caviar network. It included 4,279 paragraphs of information (over 1,000 pages) revealing electronically intercepted telephone conversations between network participants. These transcripts were used to create the overall matrix of the drug-trafficking operation's communication system throughout the course of the investigation.

Individuals falling in the surveillance net were not all participants in the trafficking operation. An initial extraction of all names appearing in the surveillance data led to the identification of 318 individuals. From this pool, 208 individuals were not implicated in the trafficking operations. Most were simply named during the many transcripts of conversations, but never detected. Others who were detected had no clear participatory role within the network (e.g., family members or legitimate entrepreneurs). The final network was thus composed of 110 participants.

Projects Siren and Togo

The data sources for the stolen-vehicle exportation (or ringing) operations were obtained within a larger investigative setting. Between 1993 and 2005, a tandem task force was united under Project CERVO.[2] Participating agencies included the Montreal Police, the Sûreté du Québec, Canada Border Services, and the Insurance Bureau of Canada (Dupont 2006: 110). The main objective of this task force was to monitor and control the exportation of stolen luxury vehicles from the Port of Montreal. Cooperation between

---

[2]CERVO is an acronym for *Contrôle de l'exportation et du récel des véhicules voles outré-mer*, which translates into the control of stolen-vehicle exportation and resale.

law-enforcement and border/insurance agencies was the unique feature, with the latter supplying documents from maritime shipping companies that contained information on suspect cargo and the identities of individuals or enterprises involved in their transportation.

Information was accessed on two ringing operations that were investigated throughout 1998 and 1999. The first operation, Project Siren, began in February 1998 when a port worker informed members of the CERVO group that a container of stolen vehicles had been recently shipped to Ghana. This shipment was subsequently seized at its transit point in Anvers, Belgium. This initial tip and action led to a close monitoring of the suspects involved in the shipment. The investigation continued for 4 months (to June 1998), during which time CERVO members monitored stolen-vehicle shipments intended for Russia, Egypt, Iraq, Italy, and Switzerland. Some vehicles were also resold in Toronto. Overall, 35 cars were retrieved according to the files that were consulted.

The second operation, Project Togo, also began in February 1998 when a Toronto-based ringing operation was dismantled and one of its participants informed the police that he was previously employed by a Montreal businessman who was also active in the resale of stolen vehicles. This initial tip was corroborated soon after by a thief who had been arrested while driving a stolen vehicle. By December 1998, the Togo investigation was under way. It spanned into February 1999 and 20 cars that were destined for France, Ghana, and local buyers in southern Quebec were retrieved.

The main data sources for both the Siren and Togo cases included transcripts revealing interrogation sessions with arrested participants and police affidavits that were used for establishing the ringing process and for assigning script designations (or roles) to participants. Physical and electronic surveillance transcripts were used to construct the ringing networks. Additional data from Royal Canadian Mounted Police intelligence reports were also compiled to corroborate relationships within the networks. Other sources of information used to verify the content of the networks were found in reports and past investigations conducted by members of the Canada Border Services who were already acquainted with some of the Siren and Togo participants as suspects in false declarations of exported merchandise.

In an initial codification of each case, 68 individuals associated to the Siren operation and 45 individuals associated to the Togo operation were identified. Not all of these individuals were actual participants in the ringing operations. All participants who simply fell into the surveillance net, but who did not have any role in the ringing process, were excluded. This resulted in the omission of 22 individuals in the Siren operation and 12 individuals in the Togo operation. The final matrices therefore consisted of a 44-node network for the Siren case and a 33-node network for the Togo case.

Operation Springtime 2001

This case deals primarily with drug distribution and collateral activities involving members of the Hells Angels Nomad chapter in Quebec and the Rockers, a smaller biker gang from Montreal. By collateral activities, I am referring to events occurring over a particular period of time surrounding Quebec's Hells Angels and their implication in the illegal drug economy, particularly within the Montreal region. More specific details on this period will be provided in Chapter 8. At this point, it is sufficient to mention that the biker organization was involved in an escalating and violent conflict with other biker gangs and illegal drug merchants during the mid-to-late 1990s. Thus, although most of the activities examined in this case study deal with drug distribution, additional activities touched on the intimidation, violence, and money management that reflected the particular context in which drug-distribution operations were taking place.

The end of the trials and the criminal justice proceedings against the Hells Angels marked the beginning of the present case study. Although the case carried the name of the final crackdown (Operation Springtime), this final ensemble of arrests was the result of a series of investigations conducted by the task-force team throughout the preceding 4 years. The task force included investigators and analysts from the Royal Canadian Mounted Police, the Sûreté du Québec, and the Montreal Police.

The first investigation carried out by this task force, Project Rush, began in 1997. Project Rush targeted the entire Hells Angels organization in Montreal, but more particularly the elite Nomad segment of the Hells Angels, their underlings (Prospects), and the ensemble of participants who were members or underlings of the affiliated gang, the Rockers. Initially, Project Rush was built on the work of a police informant who was a member of the Rockers and maintained close links with several members of the Nomads. Eventually, more informants would be incorporated into the investigation and new investigative projects would spawn from Project Rush.

The most important of these spin-off investigations was Project Ocean, which began in the fall of 2000 and was initiated when Project Rush investigators learned that one of the targeted participants was transferring money obtained from drug-distribution profits to an apartment in Montreal's east end. Project Ocean focused on this money hideout and mainly targeted the money movement and management extending from profits in the Nomads' drug-distribution activities. Investigators retrieved an ensemble of spreadsheets showing the transaction profits and costs extending from the Nomads' transactions with drug-dealers in the Montreal region. One group of names found on these spreadsheets included a drug-trafficker, who was reputed to have considerable resources for moving drug consignments and other contraband through Montreal's waterfront port, and his associate, who was responsible for carrying money from the east-end apartment to him. The money carrier eventually became an informant and revealed the drug-distribution connection between his reputed drug-trafficking associate and the Nomads.

The third case that led to Operation Springtime and included in the Hells Angels case study was Project Hammer (*Projet Marteau*, in French). Project Hammer also began in 2000, after the completion of another investigation that targeted a cocaine distribution network in the Montreal region. The previous investigation centered on a reputed drug-trafficker and revealed that his main supplier was an associate of one of the Nomads, who were under investigation in Project Rush during that time. Following the premise that "everything that leads to a Nomad is governed by the Nomads," the task force initiated Project Hammer to monitor and eventually dismantle this extensive cocaine distribution route.

The final network that is analyzed in Chapter 8 was extracted from a massive set of electronic and physical surveillance records that were submitted as evidence during the trials of 131 individuals. Media reports revealing the scope of this evidence consistently stated that the police recorded 270,000 logs of interaction bits between individuals monitored during the ensemble of investigations leading up to the March 2001 crackdown. This number is indeed a fair assessment of the initial set of files that my research team and I were dealing with when we took our first steps toward reconstructing the network surrounding the Hells Angels during this investigation. A first extraction of all logs recording nonconversational interactions (e.g., pager alerts, unanswered calls, busy signals, wrong numbers) reduced this number considerably. After three test trials on random samples of 100 logs, we found that the cut-point in file size was 400 bytes, with nonconversational interactions equal to or falling under this file size. Applying this 400-byte filter thus eliminated all telephone logs that resulted in an empty file and decreased the number of logs to 20,502. These logs recorded interactions between 1500 individuals. However, as in the previous case studies, not all individuals falling in the surveillance net were participants in the criminal network.

The final network of 174 participants is the result of a selection process that excluded all individuals who were not targeted by the police and for whom no additional evidence could be provided to illustrate their participation in the criminal operations that were under investigation. Project Rush comprised 61 of these participants, Project Ocean added 81 participants, and Project Hammer added 32 participants.

Street Gangs and Drug Distribution in Montreal North

This case study is also based on multiple investigations. Data obtained from the Montreal Police's central intelligence base, the *Automated Criminal Intelligence Information System* (ACIIS),[3] was used to reconstruct the organization of drug-distribution operations in Montreal North. These operations were targeted during three separate investigations between 2004 and 2007 by the Montreal Police, who believed that the criminal activities were under the control of one of the more reputed gangs in Montreal—the *Bo-Gars* (or Handsome Boys, in English). Because the trials extending from two of the investigations were still ongoing at the time of analysis, their names remain confidential and I simply refer to Investigations A, B, and C. Investigation A began in February 2004 and ended in April 2005, with the arrests of 27 individuals who were accused primarily of importing and distributing crack and cocaine in a Montreal North neighborhood. Investigation A was the largest of the three investigations under study and it was the only case to offer electronic surveillance information amongst the available data sources. Investigations B and C, which were direct extensions of observations made during Investigation A, both began during the fall of 2006 and ended in June 2007, with the arrests of 24 individuals who were targeted in Investigation B and 11 individuals targeted in Investigation C. Investigation B concentrated on street dealers of marijuana and crack, while Investigation C focused specifically on the activities of a group of wholesalers who were supplying some of the dealers targeted in Investigation B. Overall, 101 individuals were monitored during the investigations—45 in Investigation A, 30 in Investigation B, and 26 in Investigation C. This list of 101 individuals was used as a starting point to reconstruct the final network. This final network was comprised of 70 participants and was based on information obtained from three data sources.

Electronic surveillance transcripts were the first data source to be coded. As mentioned earlier, only Investigation A had the resources to include such a monitoring technique. That electronic surveillance data was lacking for two of the investigations was not a serious problem since the importance was to construct the network as fully as possible. It was therefore important to gather as much information as possible on Investigation A because it was the root of the other two investigations. The electronic surveillance source revealed direct connections between 39 individuals. The second data source to be coded was comprised of co-offending records that revealed who

---

[3]The ACIIS is a system that is updated and shared by law-enforcement agencies across Canada.

amongst the 101 monitored individuals had been arrested with others on the list, excluding arrests made during these three investigations. Adding these co-offending records revealed links between 26 participants in the network that were not detected with the electronic surveillance data, thus increasing the size of the network to 65 individuals. The third data source consisted of physical surveillance observations made by police investigators and patrol officers during the course of the three investigations. Five other individuals were added with this data, bringing the network to 70 individuals.

III. Designing the Criminal Networks

Once the data for each case study was accessed, gathered, and the selection of targeted participants in each case was completed, the next step was to construct the networks. This section presents the two basic components for constructing social networks, the matrix and the sociogram.

The Matrix and the Sociogram

A social network represents a set of actors that are in direct or indirect contact with each other. There are two types of social networks. Egocentric or personal networks are those that are designed around an individual (ego). Sociometric or whole networks are those that are designed around a given event or setting. The analyses produced throughout this book rely on the sociometric variety.

Whether in an egocentric or sociometric framework, social network analysis always begins with a matrix (see Fig. 2.1a) that can subsequently be visualized in a sociogram (see Fig. 2.1b). The network in Fig. 2.1 contains four nodes (N1–N4). The connections that are coded in a matrix could mean many things. They may be coded in binary format, which indicates the simple absence (coded 0) or presence (coded 1) of a relationship (e.g., Node 1 telephones Node 2). They may also be coded for values, which indicates ordinal or continuous variability within the relationship (e.g., how many times nodes telephone each other, how many times nodes have been arrested together in the past, how many times nodes frequent each other in a given time period). Finally, a matrix may also be coded for directions or symmetrical relationships. A relationship is symmetrical if it simply tells us that Node 1 and Node 2 make contact with each other, in a binary or valued format. A relationship is asymmetrical if it tells us that Node 1 makes greater contact with Node 2 than vice-versa (e.g., Node 1 called Node 2 twenty times, but Node 2 called Node 1 only once).

Fig. 2.1a A matrix

(a)

	N1	*N2*	*N3*	*N4*
N1	-	1	1	1
N2	1	-	0	0
N3	1	0	-	1
N4	1	0	1	-

Fig. 2.1b A sociogram

(b)

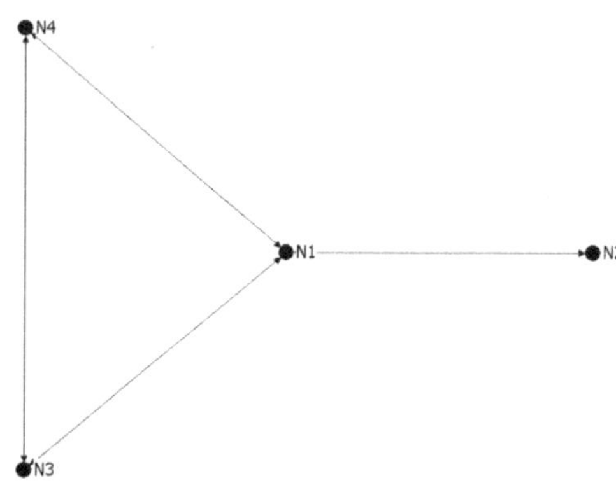

The example in Fig. 2.1 represents a binary and symmetrical matrix. We see that N1 is the only node to have contact with all other nodes. N1 forms a triad with N3 and N4. N1 also serves as a bridge between N2 and the other two nodes. In such small networks, the sociogram display is less crucial since these patterns of connectivity are visible by simply observing the matrix. However, larger networks, such as those presented throughout this book, are aided by a sociogram display.

All the case study matrices were based on the law-enforcement data sources presented in the previous section. Except for one case (the street gang network), the sources were derived primarily from electronic and physical surveillance records. Once the final list of participants in each network was completed, these sources were used to establish which participants communicated with others in the various ways. Each matrix was comprised of valued, directional data. However, the valued and directional data were only used in some specific analyses throughout the case studies. Most of the analyses were conducted with the simplified binary, nondirectional versions of these matrices.

Several of the matrices that were compiled for the subsequent case studies are provided in the book's Appendix. This last section of the book includes: (1) the valued-directional matrix for Project Ciel; (2) the valued-directional matrix for the overall network that was targeted during Project Caviar (because of its size, this matrix is divided into four parts); (3) 11 valued-directional matrices representing the composition of the Caviar network at consecutive phases during the investigation (see Chapter 6); and (4) the binary-nondirectional matrices for Projects Siren and Togo.

Assembling the Final Network Representation

Social network analysis is sometimes more reputed for the beautiful graphics that are created to visualize network data than for the substance and propositions that are fleshed out of such representations. As Collins (2008) pointed out in his own experience studying visual displays of violence in a multitude of situations: *A picture is worth a thousand words only for those who already have internalized an adequate vocabulary* (Collins 2008: 5). There is much truth to Collins' statement in that, to the untrained eye, most network representations emerge as mere chaotic displays—for my 5-year-old son, some of the sociograms in this book looked like spaghetti; my 7-year-old daughter, however, was quick to recognize the "more important" people in the various network images. To the trained eye, a sociogram triggers a sequence of focus questions: How is the overall network structured? Are relationships in the network dispersed, dense, or segmented? Is there a concentration of relationships around key nodes? Is there a chain-like quality to the network? Are the key nodes positioned as brokers or bridges between links in the chain? Does the network have a clustering pattern at the overall or localized level? If the graph is directional, is some level of reciprocity or asymmetry observable? All these and several other questions come to mind when exploring a sociogram for the first time.[4]

---

[4]Ucinet 6 (Borgatti, Everett, and Freeman 2002) was used to prepare the matrices, shape the sociograms, and generate the network analyses that appear throughout the book's case studies. Although there are several network programs that are available for achieving these objectives, the Ucinet package is simple to learn and offers an impressive range of measures and statistical procedures for analyzing networks. More recent versions of this software integrate additional programs (NetDraw, Pajek, and Mage) that facilitate the reshaping of sociograms and expand the inventory of analytical options. Ucinet is also amongst the least expensive of network software packages.

Once an initial appraisal of the network is obtained, the next step is to re-shape the sociogram in order to channel the principal analytical questions and facilitate its visualization for others. In some of the networks presented throughout the subsequent case studies, centralization was important enough to reshape the network in a way that highlighted this feature and emphasized the positioning of key participants therein. In other case studies, other features of the network were also introduced to help shape interactions in a specific way. Such features were often nonrelational and highlighted other aspects of the inherent social organization (e.g., roles, legitimate/criminal status, or organizational ranks) and crime-commission process (e.g., the script). In such cases, the sociogram was adjusted to display the features that would be introduced throughout the analyses in a given chapter.

Of course, even the trained eye requires help in managing network data and in establishing precise observations and results. Thus, in the process of preparing network data toward the final sociogram representation, we begin by assembling a matrix that will be explored in a sociogram that highlights the features under observation, but the final analysis is based on the many coefficients that have been designed for measuring various network dimensions. The principal measures used throughout this book were designed to understand how criminal networks vary in their centralization around key participants. The next section will elaborate on such measures.

IV. Centrality and Analogous Network Measures

Centrality is arguably the most popular operational concept used by social network analysts. Actor or node centrality measures tell us how the nodes within a network are positioned. The two most-common centrality measures are degree centrality and betweenness centrality, but social network analysis offers a series of distinct measures that illustrate the variety of ways in which a node could be central. In this section, I present a brief description of the measures used throughout most of the case studies in the following chapters. Aside from the two most-common centrality measures, I also introduce variations on each (eigenvector centrality and flow betweenness) as well as discussing a noncentrality, but nevertheless relevant, measure (clustering coefficient) for understanding individual positioning with a network.

Aside from the measures of actor centrality, the analyses in this book also explore the extent to which the overall networks are centralized in various forms. Each of the network centralization measures offer a contextual description of how a network gravitates around one or a few nodes that are central in the various ways revealed by the array of measures. Actor centrality rankings and scores must be considered in the wider context of whole network centralization in order to be fully appreciated. All network centralization measures may range from 0 to 100%. Thus, a node that has the highest degree centrality in a network that is high in degree centralization is substantively different from a node that has the highest degree centrality in a network that is low in degree centralization.

In the Thick of Things

Degree centrality is the simplest of the centrality measures. It is a straightforward measure of the number of direct contacts surrounding a node. Nodes with high degree centrality are therefore those that attract a high concentration of direct connectivity within a network. Degree centrality scores are generally presented as percentages of the overall number of other nodes, thus, a node with ten direct contacts in a 20-node network will have a degree centrality score of 10/(20 – 1), equaling 53%. Expressed as such, degree centrality scores vary between 0 and 100%, with 0% indicating that the node is an isolate and 100% indicating that the node is in direct contact with all other nodes in the network.

In Between

Betweenness centrality extends from degree centrality and, like other alternative centrality measures, introduces the nuance that it is not the quantity but the quality of connections that is important. This measure incorporates the indirect contacts that surround a node. Betweenness centrality measures the extent to which a node mediates relationships between other nodes by its position along the geodesics within the network. A geodesic is the shortest path (or number of degrees) connecting a dyad (a pair of nodes). The greater a node is located along the geodesics in the network, the higher its betweenness centrality. This measure essentially represents the ability of some nodes to control the flow of connectivity (or communication) within a network. Controlling the flow within the network in this indirect manner is the broker's edge. The index for betweenness centrality for a given node is equal to the proportion of times that that node is positioned along the geodesics between dyads. The maximum number of geodesics is equal to the number of dyads not including the node or $(g - 1)(g - 2)/2$, where g is equal to the total number of nodes in the network (Wasserman and Faust 1994). Thus, in a five-node network, each node, in theory, will be able to mediate between six dyads. The minimum score is 0, which means that a node falls on no geodesic. The maximum results when a node falls on all geodesics (the star graph or pure broker configuration).

In the Thick of the Thick

Eigenvector centrality extends from degree centrality by resolving an important puzzle inherent in the simpler measure: are all actors with the same degree centrality equally central? The answer is no, and the reason is that it depends on how an actor's contacts are themselves connected. Eigenvector centrality adds a qualitative aspect to the number of direct contacts that make up a node's degree centrality. Developed by Bonacich (1972), it begins with the assumption that some nodes are central because they have a high degree of direct contacts and because these direct contacts are themselves in direct contact with a high degree of nodes in the same network. In short, eigenvector centrality measures the extent to which a node is connected to other nodes that are high in degree centrality in the network. Bonacich was explicit that such centrality does not increase an actor's power within a network because high eigenvector centrality is an indication of actors who are in contact with others who are not dependent on them because they have their own high degree centrality.

Less Efficiently In Between

Developed by Freeman, Borgatti, and White (1991), flow betweenness centrality widens the scope of mediation measured by betweenness centrality. Whereas betweenness centrality is based exclusively on the shortest paths (the geodesics) between nodes in a network, flow betweenness centrality allows us to loosen such restrictions by considering all possible paths between nodes. Because it does not rely on the shortest paths between nodes in a network, it is a measure of "less-efficient brokering." The main condition in this measure is that when calculating the extent to which a node is positioned along all paths uniting all other nodes, no node could be "visited" more than once. Individual scores are generally presented as a ratio between (or standardization of) a node's position in the flow of connectivity between other actors over the extent of connectivity in which the node is not an intermediary.

Localized Clustering

Although not typically included amongst the set of centrality measures, another way of accounting for individual positioning within a network is by measuring the extent to which a node's direct contacts are clustered. This measure, the clustering coefficient, establishes whether a node is part of a localized cluster or clique within the overall network. The clustering coefficient accounts for direct connections between a node's direct contacts. It is essentially a measure of personal network density within a whole network. The relevance of this measure depends on the overall structure of a network. Dense networks are composed of cliques and a higher degree of connectivity between nodes. Dispersed or chain-like networks in which nodes are minimally connected to each other are low in density. The clustering coefficient measure was proposed by Watts (2003, 1999) to be more suitable for illustrating the extent to which some low density networks may be composed of dense segments (clusters). This pattern was expected to fit large networks in which high overall density would be difficult to achieve, but within which nodes would interact in smaller, tightly knit subsets within the whole. The clustering coefficient for a complete network is the average of clustering coefficient scores for all nodes. The higher the average score, the more individual nodes have personal networks within the whole that are densely connected and the more likely the whole network is to be comprised of clusters of localized cliques—in a network that is already dense to begin with, this is obvious; the clustering coefficient is a revealing measure particularly in low density networks.

Note that some traditional centrality measures, such as closeness centrality or reach centrality, were not presented in this section. The omission of these and other measures was based primarily on my failure to observe their applicability in the analyses presented throughout this book. Other measures, such as Gould and Fernandez's brokerage leverage, will be used in later chapters, but because their applications are restricted to single analyses, they will be presented only at that point.

V. Challenges in Criminal Network Analysis

Probably the main strength of social network analysis is that it allows us to identify and scrutinize structure in areas where we are quick to assume structure. This is particularly useful in criminology where, on one hand, mainstream researchers tend to deny or underplay the level of organization needed for crime (Gottfredson and Hirschi 1990) and, on the other hand, some researchers of organized crime have assumed the other extreme by overstating the corporate or bureaucratic organizational form as the traditional framework for understanding serious crime. Within a network framework, the structure of the network is sought after empirically and low-level or high-level organizational forms may emerge from the analysis.

The following sections elaborate the main limits associated with social network analysis in the context of the law-enforcement data sources used throughout the book. All problems presented address the simplest binary, nondirective matrix—for directive and valued matrices, these limits are certainly heightened.

Clarity and Attributes in Relational Data

Network data is relational in that it exposes the details concerning relationships or interactions between two or more actors converging in the same general context. Such data is fundamentally different from the typical attribute data used in the social sciences. However, some of the most important limits concerning such relational data are due to problems associated with the attributes of the network actors. This is particularly the case when using law-enforcement surveillance data as the principal source for constructing a network.

A consistent problem that arises when coding electronic or physical surveillance data concerns the clarity of the conversation and the identity of the detected participants. In the case of ongoing investigations, many conversations are unclear for several reasons and this often results in difficulties in identifying one or both participants taking part in the conversation. New individuals who are unknown to the investigators arise on occasion. In electronic surveillance, the identification of voices is a particular challenge. Finally, the problem of aliases used by network participants could also present important difficulties when reconstructing a network based on data retrieved from police investigations. While these problems were indeed recurrent across investigations, they were not as overwhelming as suggested by critics of such data and method.

The first thing that critics of law-enforcement data must understand is that in as much as criminal-network participants may attempt to conceal their identities and actions by remaining as discreet as possible or by inventing private codes to conceal the content of their conversations, it remains that such attempts to conceal also hinder communication between the network participants themselves. In this sense, and contrary to popular opinion, clarity in the criminal network is as (or more) important than concealment. The experience of a former international cannabis smuggler (Howard Marks) shows us why the often-repeated criticism that law-enforcement surveillance data is largely tapping into hidden and sophisticated codes is unfounded: *International dope smugglers have to make thousands of phone calls. There are many who say they never use the phone because it's too insecure. They are either lying or not doing any business. Dope smuggling is fraught with unexpected obstacles. Problems have to be solved quickly. The multinational and multicultural nature of the personnel involved severely limits the possibility of utilizing any workable encryption of the intended content of the phone call. All dealers and smugglers use simple and fairly transparent codes. Any attempt at sophisticated coding quickly leads to disastrous misunderstandings. I have never heard or made a dope-smuggling call which isn't obviously just that* (Marks 1997: 306–307).

Marks' experience communicating inside criminal networks is closer to my experience studying such networks. While the ensemble of surveillance logs gathered throughout an investigation contains a substantial amount of observations that expose nothing more than mundane and noncriminal daily routines, those that expose criminal content are generally clear on what is being discussed. When some level of concealment was attempted, investigators had enough resources to resolve voice detection and identify individuals operating under an alias. Thus, when someone starts a conversation with the typical *Hi, it's me*, investigators were typically able to identify who this person was through the telephone number and previous voice matches, and although it was common for participants to use multiple telephones within short periods (a pattern which was simple for investigators to keep track of), it was not common for them to exchange telephones with each other. In cases where individual referrals were disguised on a consistent basis through nicknames as sophisticated as the *Big Guy*, the *Tall Guy*, the *Bald Guy*, or the *Other Guy*, even police investigators were able to resolve such identity puzzles over an extended period. Indeed, time was the most important aspect to arrive at a clear vision of the criminal network under surveillance. Aside from the resources that were at the disposition of investigating officers and the scope of the police mandate, the duration of the investigation was the main resource for unraveling ambiguities within the network. Thus, networks that contained some level of identity and conversational concealment were generally monitored for longer periods until these matters were clarified. The longest and most complex cases (Caviar and Operation Springtime) spanned more than 2 years.

I was obliged to clarify participants' identities in only three instances across all the networks studied in this book. All three occurred in the Project Caviar case. In all three instances, the participant's identity was found by matching the alias' network relationships with that of an already identified participant—in all three cases, the match was subsequently confirmed through additional conversational data. This strategy is consistent with the standard use of structural equivalence for resolving missing data problems in general social network research.

Most network designs are faced with a missing data problem. This is particularly the case with nonsurvey research, such as this book's law-enforcement-based case studies. Regardless of their resources, law-enforcement and other criminal justice agents generally have a partial vision of a criminal network. In terms of relational data, missing data comes in two forms—nodes may be missing or relationships between established nodes may be missing. These limits may be addressed from two angles: missing data beyond and within the parameters of the final network representation.

Missing Data Beyond the Final Network Representation

Limits concerning missing data beyond the final network representation are often formulated as the boundary specification problem (Laumann, Marsden, and Prensky 1983). Once the parameters of a setting are clearly established, it is often a challenge to assess the extent to which individuals interacting within those parameters are missing. Thus, what social network researchers refer to as a "whole network" is rarely whole to begin with.

The accuracy of data compiled from criminal justice sources depends on the criminal justice stage from which they were extracted. The least accurate of sources would be that obtained from general law-enforcement monitoring of a given criminal operation beyond the specific targets under investigation. Relying uncritically on this latter depiction of a criminal network would expose us to everyone who fell into the law-enforcement surveillance net, regardless of their participation in the criminal operation under investigation—in the case studies in this book, those who fell in the net included everyone from family members (and friends of family members) to pizza deliverymen. Because the data obtained from each investigation generally included all surveillance transcripts, the case studies generally began at this outer segment, however, each of the monitored networks were filtered so that individuals falling into the net for a variety of reasons unrelated to the criminal operations were excluded from the analyses. In this book's case studies, the parameters (the context) of the networks under analysis were therefore defined in accordance with the targeting range of the law-enforcement investigations. Figure 2.2 displays how criminal network scope and precision are inversely related when using sources from various stages within the criminal justice system.

The most-accurate information that may be compiled from criminal justice data is that confirmed by a guilty verdict (at the center of Fig. 2.2). Such precision decreases as we extend from the final court verdict and move toward data based on accusations, arrests, and targeting during an investigation. Although offering the most certitude, reconstructing a criminal network based exclusively on participants who were found guilty will generally result in a representation with only a small number of participants. Thus, any researcher working with criminal justice data must weigh precision against such a restricted vision. Table 2.1 displays the trade-off for the networks studied across this book. Once again, the final networks analyzed in

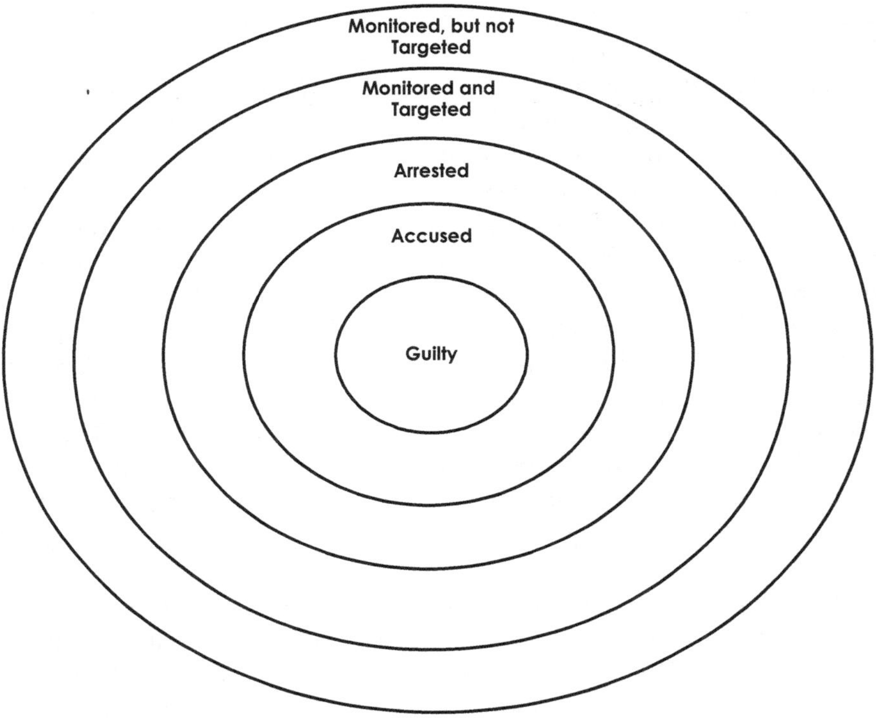

Fig. 2.2 The scope of criminal networks within criminal justice circles

Table 2.1 Number of participants found per case across criminal justice stages

	Ciel	Caviar	Siren	Togo	Gangs	H.A.
Monitored	75	318	68	45	101	1,500
Targeted	25	110	44	33	70	174
Arrested	11	25	14	12	44	138
Accused	8	22	1	2	–	131
Guilty	5	14	1	2	–	131

subsequent chapters were based on participants who were targeted and not on all individuals who were monitored. The drop in network size between the monitored and targeted stages is quite significant, particularly for the drug-distribution networks. The most striking drop (almost 90%) took place within the Hells Angels case (Operation Springtime). In the Ciel and Caviar networks, about two-thirds of those monitored were filtered from the final analysis. In the investigation targeting the gang-based drug operations, the drop was roughly 30%. This was similar to the Siren and Togo cases, in which about a third of the people were filtered when moving from the monitored to targeted vision of the network.

Once again, the main reason for such differences has more to do with the length of the investigation and the level of resources that investigators had at hand than the type of criminal activities in operation within the networks. Longer and more resourceful investigations will likely net more people and many of these people may simply be part of the wider segment of (noncriminal) network extensions of the targeted participants—such individuals are thus proximate to the criminal network, but they are not participants.

Shifting from the targeting stage of the law-enforcement investigation to arrests also led to important drops in network size. The most important occurred in the Caviar network, with a drop of 77% when restricting the focus from participants targeted to participants who were arrested. The number of participants in the Siren and Togo networks decreased, respectively, by 68 and 64%. Project Ciel decreased by 56% of its participants. Lesser drops occurred in the gang-based network (37%) and Operation Springtime (21%).[5]

The number of participants continues to decrease as we enter the accusation and conviction phases of the criminal justice system. In some cases, the idea of a criminal network becomes largely irrelevant. The percentage of arrestees who were accused varied across cases, with the Project Ciel, Project Caviar, and Operation Springtime cases containing high proportions and the Projects Siren and Togo cases resulting in only one or two accusations.

In general, those found guilty or accused for their participation in the criminal operations were positioned at the core of each network. Thus, a choice must be made on whether the sought after criminal network representation should incorporate the periphery to varying degrees. The further we shift from the guilty verdict and make our way into the extensive targeting of individuals surrounding a criminal venture, the more we are likely to tap into the periphery. Basing a criminal network representation on only

---

[5]Note that the street gang and Operation Springtime investigations were the only two that applied gangsterism legislation as a basis for arrests and accusations. Gangsterism is the Canadian version of the American RICO legislation. Individuals found guilty of such a crime are those who are proven to be part of an ongoing criminal organization (or *gang* in French), defined by article 467.1 of the Canadian Criminal Code as: *a group, however organized, that (a) is composed of three or more persons in or outside Canada; and (b) has as one of its main purposes or main activities the facilitation or commission of one or more serious offences that, if committed, would likely result in the direct or indirect receipt of a material benefit, including a financial benefit, by the group or by any persons who constitute the group.* Unlike traditional predicate crimes, gangsterism is a crime of association and such direct or indirect association is likely to increase the likelihood of arrest for a wider range of criminal-network participants.

the most valid of criminal justice sources will likely omit such an important, yet discrete, segment of the network. In this sense, data offered at the stage of the law-enforcement investigation gave a better trade-off than data offered and validated within the prosecutorial phase. Reconstructing the network at the targeting stage of the investigation resolves much of the missing node problem that would be associated with the more precise visions offered by the prosecutorial phase, while also avoiding the inclusion of many non-participants who would be detected at the monitoring stage of the police investigation.

Missing Data Within the Final Network Representation

Once a final set of nodes is selected for inclusion in a network, limits also emerge in regard to the accuracy of the relational data linking nodes within a network. Here, the problem is not missing nodes, but missing links or edges. Table 2.2 offers a general idea of the extent of connectivity recorded for each network. Much of the missing edge problem could be addressed by assessing the density of the various networks and comparing them with results from past research in this area.

Density is a measure used to assess the extent to which nodes in a network are connected to each other. The measure is a simple proportion of the number of ties (or dyads) observed over the maximum number of ties that would have been possible in the network. In a 25-node network such as Project Ciel, the maximum number of possible ties is 300: $(25 \times 24)/2$. Since only 35 dyads were recorded with the data used to create the networks, the Ciel network's density is 11.7% (35/300). When converted into a percentage, a network's density ranges from 0 (no ties between any nodes, thus, the absence of a network) to 100% (all nodes are tied to all other nodes, thus a clique network).

All the networks designed for this book were relatively low in density. The Ciel network was the densest of the networks examined. Typical of larger networks, density was lowest for Project Caviar (3.4%) and the Hells

Table 2.2 Relational features of final (targeted) case study networks

	Ciel	Caviar	Siren	Togo	Gangs	H.A.
Number of dyads	35	205	103	47	237	800
Max. dyads possible	300	5,995	946	528	2,415	15,051
Density	11.7%	3.4%	10.9%	8.9%	9.8%	5.3%

Note: Results are based on binary, nondirectional matrices.

Angels network in the Springtime Operation case (5.3%). Such low density does not appear to be a mere product of the law-enforcement data at the source of the various network representations. We should expect density to be low in covert activities. Past research supports this expectation and the current findings. Research on street-gang networks, for example, has repeatedly described such settings to be dispersed and low in cohesion (Klein and Crawford 1967; McGloin 2005; Papachristos 2006; Klein and Maxson 2006). Amongst the few studies that examined criminal networks, low density was also a steady result (Finckenauer and Waring 1998; Natarajan 2006). One slight exception is Baker and Faulkner's (1993) research that had networks with density ranging from 23 to 35%. While these latter results would still indicate low density in conventional social network research, it is also important to note that although the data sources used for the Baker and Faulkner study were not derived from law-enforcement sources, the three networks under examination were relatively small in size and comprised of deviant operations extending from a legitimate work environment—both of these features would have an impact on increasing network density.

Another feature that must be considered when assessing the findings from this book's case studies in view of data error due to missing ties between nodes in the networks concerns the robustness of the principal measures used throughout the analyses. The analyses center primarily on the identification of various types of key participants in the criminal networks under study and the centrality measures that are used to identify such key participants have proven to be robust measures in view of missing data problems (Borgatti, Carley, and Krackhardt 2006). More importantly, the analyses focus particularly on centrality rankings rather than centrality scores—the former are even less influenced by changes in the network's composition than the already robust latter.

Are Central Participants Simply Central Targets of a Police Investigation?

There is another criticism that is often made in regard to the search for key participants in criminal networks that are based on law-enforcement data. In the context of their investigations, law-enforcement agents generally begin by targeting one or two individuals and gradually expand toward other core or peripheral participants in the monitored criminal operation. The problem when using data from such investigations for network analysis, and particularly for the key player issue, is that the participant who appears to be central in a criminal network is often the person who was central (or the starting point) in the investigation (Sparrow 1991). Indeed, this may be the case, but it remains far from given. In as much as the initial target of an investigation will generally be central in terms of degree (direct connectivity), it is the emergence of other central participants who were not detected as key players during the investigation that becomes the highlight of the analysis—this generally occurs when we move beyond degree centrality. At the same time, the extent to which a central target of an investigation is, by definition, a central participant in the network depends on the length of the investigation, the organizational structure of the network, and the ability of the investigation to tap into this organizational structure over time. Longer investigations, once again, will likely allow investigators to see and potentially learn more than initially expected. Of course, whether the investigators themselves acknowledge the presence and rise of other central participants is another matter. In a network analysis of law-enforcement data, it is not the investigators' understanding of the network that is assessed, but the observations that are made throughout the investigation. As the case studies will demonstrate, there are often striking differences between investigators' interpretations of a criminal operation and the findings that emerge from a network analysis of their observations during the investigation.

In other cases, investigators may have less control over what they see. A network participant may be the initial target of an investigation, but this participant may also be positioned strategically enough to avoid such direct visibility on his or her activities. At the early stages of an investigation, this participant is the central node in the network simply because all "eyes" are set on him/her. However, as the investigation continues, the organizational structure of the network unfolds and the strategic positioning of this participant emerges, leading to a displacement in higher (degree) centrality toward other, less-targeted participants. Although the data renders this process observable, it is rarely accounted for in law-enforcement investigations because electronic and physical surveillance data are more often used for their anecdotal than aggregate value. In other words, investigators are more intent on seeking the incriminating conversational piece or physical encounter than assembling a more complete representation of who interacts with whom in the targeted network. The variability of central targets and central participants across an investigation are examined more fully in Chapter 6. The strategic positioning of some central targets is illustrated in Chapters 8 and 9.

Are Law-Enforcement Intercepted Networks Simply Failed Criminal Networks?

While the use of law-enforcement cases that ended in the arrests of some criminal network members creates a selection bias based on the "failed" criminal operation, we must not forget that one criticism of the social network field is that many analyses are based on successful cases of firms and their efficient economic pursuits (Podolny and Page 1998). Thus, whereas research on conventional social networks has relied on the successful case and largely avoided the failed case, criminal network research, because of data availability restrictions, is largely dependent on the failed case. But are the criminal networks studied in this book simply failed cases and do they show us nothing beyond this specific outcome? Although I do accept that failure is a feature of each, I do not maintain that this was due to the structure of any one of these networks. All of these networks were targeted not because of a fault in their structure, but because they became the central focus of police preoccupations during a specific time frame or because of an initial tip made to a previously uninformed group of law-enforcement agents who decided to follow the lead once it was made. The tip in itself is a sign of criminal network vulnerability, but this risk confronts all criminal coparticipation structures. Law-enforcement agents become strategic and disruptive only after a criminal network becomes overexposed or through attention from an outside source.

Thus, the criminal networks that will be examined in diverse ways across the following chapters should not be seen as networks that were destined to fail to begin with; the criminal networks failed because they fell within the scope of persistent law-enforcement targeting. There is much to be observed from law-enforcement data on criminal networks that reveal individual and collective behavior before the ultimate fall (arrests). Indeed, that failure was imminent in all cases allows the identification and emergence of resilience and adaptation patterns. This latter feature would not necessarily be foreseeable if a criminal network was not under the significant control of external forces.

Chapter 3
Partnership Configurations in Illegal Drug-Importation

A consistent finding in criminological research is that the majority of crime events involve the collaboration of two or more people. This has been the principal fact at the center of co-offending research. Reiss (1986) was one of the first to look at the co-offending phenomenon from new perspectives. One twist that he offered concerned the focus of analysis. Whereas an assessment of criminal events (or arrests) confirm that half of crimes are committed by more than one person, a focus on individual offenders and their crime-commission patterns heighten the co-offending phenomenon even further. Using juvenile crime event (or arrest) data on burglaries and robberies, Reiss found that half of burglaries were committed by two or more offenders. The co-offending figure increased to 67% if the proportion of offenders who committed burglaries with two or more people was taken into consideration. Findings on robbery were also consistent. Just under half of robberies were committed by two or more offenders, while three-quarters of robbers committed their robberies with others (Reiss 1986: 124–125). Reiss also found that 17% of offenders always committed crimes with co-offenders, while 63% varied between lone and co-offending across their crimes (Reiss 1986: 125).

In regard to the duration and stability of co-offending relationships, Reiss was consistent with Sarnecki's past (1986) and more recent (2001) research on co-offending patterns in Sweden—that such relationships, particularly amongst juveniles, were generally short-lived. The instability of co-offending relationships was also a general pattern found by McGloin, Sullivan, Piquero, and Bacon (2008) in their research on a sample of juvenile offenders from Philadelphia. However, this pattern varied in accordance with the subsample under analysis and with the offending frequency and the size of co-offending groups. In the overall sample, offenders with a larger number of arrests and those who participated in larger co-offending groups were more likely to "reuse" previous co-offending relationships. These patterns were inversed when examining the same effects in the subsample of juveniles who did reuse their co-offending relationships—here, more active

C. Morselli, *Inside Criminal Networks*, DOI 10.1007/978-0-387-09526-4_3,

offenders and those who offended in larger accomplice groups were less stable in their co-offending patterns.

McGloin et al. were correct in assuming that co-offending stability was largely a product of the limits in the pool of potential accomplices for offenders who are more active and involved in larger groups—the greater the number of offenses and the larger the co-offending group, the more likely past co-offenders will cross paths as repeat accomplices. But, in as much as co-offending is clearly revealed in official crime data, the phenomenon is even greater if we widen the scope beyond the arrest incident. This was Tremblay's (1993) main point. He argued that there was more to co-offending than the number of individuals who are arrested together for the same crime. Tremblay stressed that much crime is dependent on the general availability of offenders and that the co-offending concept should be expanded *not only to the subset of an offender's pool of accomplices but rather to all those other offenders he must rely on before, during, and after the crime event in order to make the contemplated crime possible or worthwhile* (p. 20). What Tremblay was referring to was the extent to which offenders are able to access necessary resources in a criminal network to varying degrees.

I. Resource-Sharing in Crime

For some reason, research on co-offending and research on organized crime or criminal markets have not crossed paths as much as they should. The latter line of research has been concerned with the social organization of groups, enterprises, and criminal organizations, but few have built upon the basic aggregate-level facts provided by co-offending researchers. Take Reiss' findings on predatory offences, for example, and adapt them to the criminal market settings that have been of concern to organized crime researchers. We should expect co-offending to be higher within the context of market crimes, which are transactional by definition. The proportion of co-offending in drug-dealing, for example, should be higher than that found for robbery and burglary events. Drug-dealers should also have a higher proportion of co-offending experiences in their past arrests. Such offenders should also have more consistent co-offending experiences than the 17% experienced by burglars and robbers in Reiss' data set. If we widen the scope into Tremblay's co-offending framework, the claim that any criminal market offence could be committed by a single individual becomes a contradiction in itself. Even more so than all other forms of crime, criminal market offending requires an ability to collaborate with others.

That collaboration amongst offenders is a necessary condition for crime, and particularly market crimes, is indeed an obvious statement. However, this evident condition has rarely been followed through. Instead, past research has often surpassed the co-offending condition and sought, instead, various levels of sophistication in the social organization of crime. To say that a venture to import illegal drugs is simply an illustration of offenders coming together to execute a crime in the quickest and safest way possible offers less in dramatic effect when contrasted to competing claims that the same venture is coordinated within the tightly regulated confines of a reputed criminal organization.

Indeed, mere partnerships have generally received less of a spotlight than the dominant criminal organization. However, the ability for individual illegal entrepreneurs to overlap their ventures in a multitude of partnerships offers the wide-scoped venturing that permits simultaneous participation in diverse criminal markets. This was one of Haller's (1990) main arguments in his analysis of American crime groups of the twentieth century. Partners in crime share risks and profits in joint business ventures. Diverse forms of expertise are carried into the venture from the various partners. Some may bring political protection to the mix, while others may bring financial investments, underworld and upperworld contacts, managing acumen, and trafficking or other criminal skills. Such resource sharing is the basic incentive underlying the criminal networks that offer the pool of accomplices that are necessary for most crime.

Haller's resource-sharing model is consistent with a number of studies on different levels of illegal drug-trafficking. At the street or retail level, Jacobs (1999) found a crack dealing context in St. Louis during the mid-1990s that was filled with freelancers and shifting business relationships. Eck and Gersh (2000) found that a "cottage industry" model was the most-accurate representation of the drug-trafficking trade in the Washington-Baltimore area between 1995 and 1997. Such a model was consistent with partnership configurations in that it was marked by small groups, easy entry/exit, no central control by any individual or specific group, weak organizational structures, minimal established leaders, an absence of specialization, and fluid group membership.

A similar model also reflected Hoffer's (2006) detailed research of a street-level heroin dealing network in Denver's Larimer district during the height of the "cleanup era" in the 1990s. Hoffer followed the experiences of two heroin users, Kurt and Danny, who become key dealers in the network that was heavily embedded in heroin consumption within this particular area. He provided a clear description of the transitory nature of this dealing network: *partnerships between members of this group were common. For example, in a typical three-month period Jerry and David might partner up one month, the next month David might partner up with Kurt, and the month after that Kurt would partner with Jerry. With a relatively small group, eventually everyone got to know everyone else through this shifting partnership process* (Hoffer 2006: 24). The partnership that emerged between Kurt and Danny became the central object of Hoffer's study. After local politicians and police proceeded to crackdown in the Larimer area, the heroin market dried up and it became increasingly difficult for the many users to locate reliable suppliers. Kurt and Danny combined their resources and filled the void. Each needed the other for the enterprise to flourish, which it did for about 5 years. Kurt had excellent contacts with local users. Danny had excellent contacts with suppliers. Furthermore, because police were concentrating on looking for tighter and larger criminal organizations of immigrant dealers, the local and discrete partnership that was in place between Kurt and Danny remained far from the scope of targeting. In time, the partnership became even more insulated as brokers were added between Kurt and the growing number of customers. The partnership came to an end when ambitions grew and Kurt and Danny attempted to expand through a franchise system. The failure of this system is another lesson in the limits to growth facing most criminal trade operators who have ambitions to expand beyond the mere partnership (Reuter 1983; Tremblay, Cusson, and Morselli 1998).

Partnerships are not restricted to the street or retail level. Block (1979) found partnerships, which he referred to as "combinations," at retail, franchise, wholesale, and importation levels of the cocaine trade in 1920s New York City. Adler (1985/1993) also observed such loose collaborative ventures in her ethnography of an illegal drug smuggling "community" in southern California during the 1970s. Reuter and Haaga (1989) conducted interviews with incarcerated high-level traffickers in five American correctional institutions and similarly emphasized the *small partnerships in which each partner is also involved in trading on his own account* (p. 40). Pearson and Hobbs (2001) interviewed criminal justice agents and inmates who had been active in middle-level drug distribution of cocaine, heroin, cannabis, and ecstasy markets throughout the United Kingdom. Although not dismissing it completely, they found the monolithic hierarchy model to be a limited representation and concurred with previous research that it was more *useful to think of drug trafficking as partnerships between independent traders* (p. 12). Most recently, and within Canadian and Quebec contexts, Desroches (2005) provided us with a number of examples from the experiences of 70 incarcerated importers, wholesalers, and manufacturers of illicit drugs. In almost 30% of the cases in this sample, respondents reported having been active in partnerships in their trafficking ventures. Some of these cases were 50/50 ventures that involved the long-term collaboration between participants in the partnership. Others were less cohesive and involved independent traffickers coming together for the simple advantages of resource pooling.

Co-offending, once again, is the necessary condition, but it does not have to take a form beyond the basic short or long-term partnership. The most important point to be retained from past research is that in as much as hierarchical and large criminal enterprises may sometimes be identified in criminal market settings, the presence of such governance structures are not necessary and flexibility is generally a more common and better option for crime.

In this chapter, further support is added to the basic partnership model with an analysis of a small drug-importation network. This case study illustrates the inner workings of the liquid hashish importation network that was targeted during the Project Ciel investigation in the mid-1990s. The investigation began as an offshoot of an earlier case in which officials from the Royal Canadian Mounted Police were monitoring the actions of a reputed Montreal drug-trafficker. This earlier case focused on a suspected importation operation that involved the shipping of campers with hidden consignments of hashish from France to Quebec. While the investigators of this particular camper scam did not gather enough evidence to proceed with any form of intervention, it did offer a new lead. Project Ciel was the result of the investigators' growing awareness that the traffickers in their monitoring scope were not importing hashish-filled campers from France, but were instead operating a hashish importation channel between Jamaica and Montreal.

Typical of many Canadian investigations of drug-smuggling and trafficking, the operations in Project Ciel were described as taking place within a tightly governed organization. Reports from investigators maintained that their main target (Node 1, or N1, in the network) was the "criminal organization's boss," who was described as having a firm control over his main "lieutenant" (N2) and a series of other subordinates. The investigation produced three separate seizures, with two taking place at Mirabel airport near Montreal and another occurring at Sangster airport in Jamaica.

II. Two Networks in One

Drug-trafficking operations such as those found during the Project Ciel investigation could easily be presented as a hierarchy or a looser partnership configuration. This decision is left largely to the (law-enforcement or scholarly) investigator's discretion, but a closer analysis of the inner workings of this trafficking venture does help weigh the decision in favor of one configuration over the other. An initial visualization of the Ciel network immediately points out that although the investigators' assessment of the drug-importation network maintained that the importations were tightly coordinated by N1, the network was clearly centered on two participants. Figures 3.1 and 3.2 offer

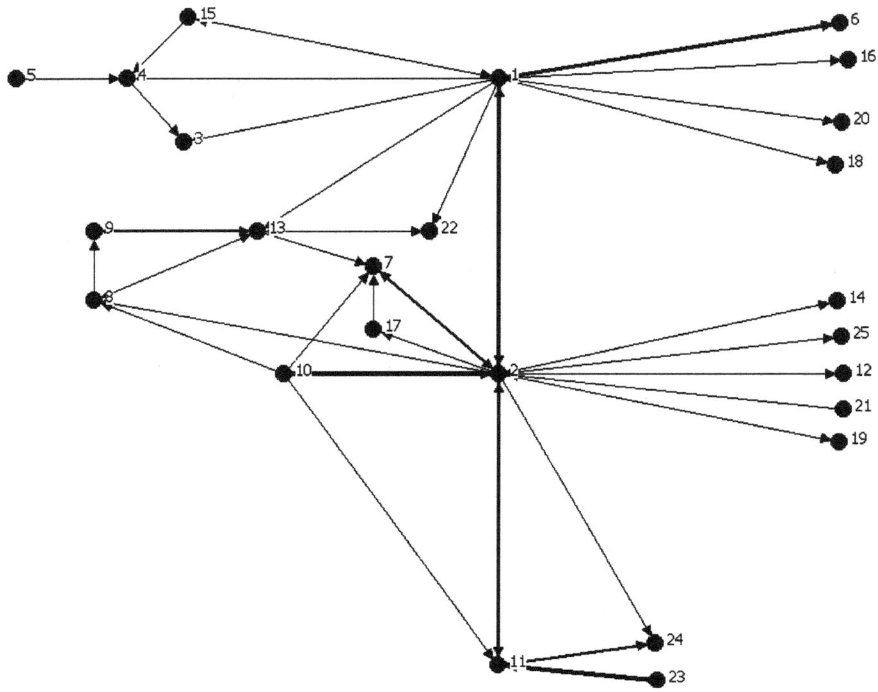

Fig. 3.1 The Ciel network as a hierarchy

two distinct interpretations of the Ciel network. The first presents a hierarchy model and the second presents a partnership model. In both figures, the darkened lines represent the more active or main communication channels between the 25 participants that comprised this network. [1]

At first glance, both figures appear valid. In the hierarchy model, N1 is positioned as the ring leader or, using Dorn, Oette, and White's (1998) term, the "cut out." This was the principal depiction maintained in the police interpretation of this drug-importation ring. N1 distanced himself from operational matters by delegating coordination to N2, who was presented as N1's lieutenant in the police reports. N2 was subsequently in contact with N11, who was the main organizer of the drug mules who were carrying the hashish across the border. These mules, N23 and N24, were N11's son and daughter. The operational risks were taken primarily within this set of mules. The remaining key link was between N2 and N10, who was incarcerated at Donnacona prison near Quebec City during the scope of the investigation and the drug-importation operations.

[1]The valued and directive matrices will only be used for these illustrative purposes. Later centrality analyses will rely on the binary, nondirective matrices.

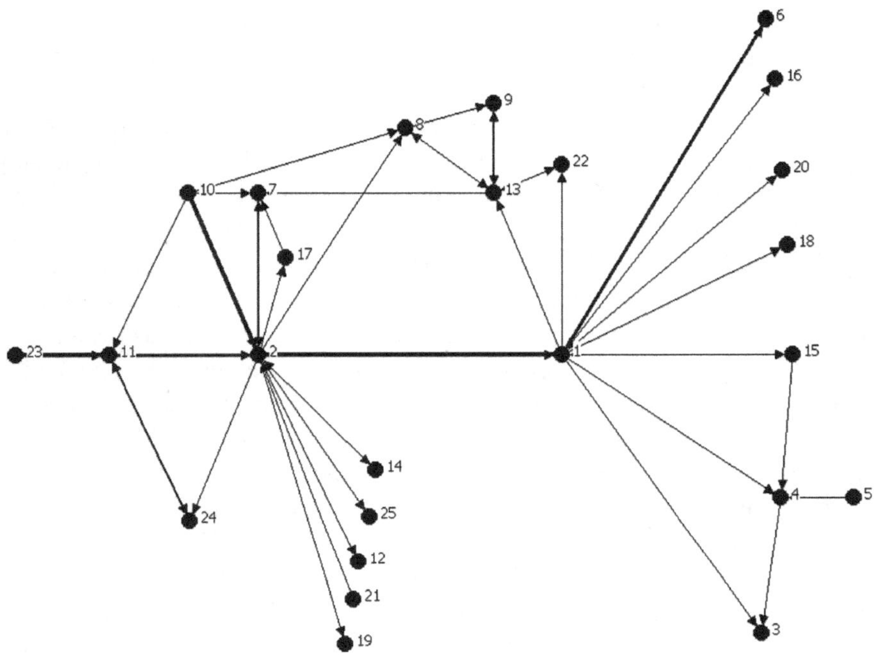

Fig. 3.2 The Ciel network as a partnership

When we analyze the Ciel network and its main channels of communication within the Fig. 3.1 representation, the makings of a hierarchy are observable: N1 delegated to N2; N2 dispatched to N11; N11 took care of the drug mules. N1, as ring leader or boss, was insulated from the most likely targeted action that was executed by these border crossers. The hierarchy model centered on N1's leadership does have some initial support, but some questions do emerge to contest this interpretation. For example, why was someone in prison so implicated in this drug-importation operation? More details regarding N10 revealed that he was amongst the most highly reputed drug-traffickers in Quebec and had been implicated and incarcerated for his participation with established criminal groups in the Montreal region from as far back as the 1970s. Of course, a case may be made that, based on his reputation, N10 was the true leader of this hierarchy and regardless of his inmate status, remained powerful enough to govern a drug-importation ring from such a confined position. But his scattered involvement with past groups revealed a different pattern since it would be less conceivable that he was the leader of all operations with all past criminal groups, most having reputed leaders of their own. In the Ciel case, investigators reported that it was N10 who brought N11 into the network by introducing her to N2. By creating this link between a coordinator and a mule in the network, N10 continued to preserve a stake in these importations even though he was sitting in a federal penitentiary. N10 brought a key resource into the network and such resource sharing is the basis of the next organizational model within which the Ciel network could be refitted.

Figure 3.2 is based on exactly the same communication network as Fig. 3.1. In Fig. 3.2, the Ciel network was redesigned to fit the partnership model. The network was centered primarily on the resource sharing of N1 and N2. Thus, in this representation, central participants are identified, but any form of hierarchically induced order and command structure is omitted from the analysis. We already know that N10 connected N2 and N11, but N2 contributed much more considerably to the network. N2 was responsible for what is referred to as the "action segment" of the network. This action segment incorporates all relationships that revolve around the movement of drugs across borders in the most efficient manner possible. N2 kept track of the actions of the drug mules (N11, N23, and N24) and was consistently informed by a group of employees and patrons who transferred messages to him from his Montreal bar (N7, N12, N14, N17, N19, N21, N25). N1 remained a key player, but in this configuration, he had a "hands on" role. He brought security to the partnership through his contacts with upperworld actors, such as an airport official (N6) and another contact who lent his name for money transfers (N16). While the links with N3, N4, N5, and N15 were retained as part of the Ciel network, a closer analysis shows that this particular segment was part of another operation that N1 was partnering beyond his resource sharing with N2. Finally, both N1 and N2 were in direct contact with buyers (N8, N13) who were waiting to distribute the hashish that was carried into the Montreal region through this importation network.

This partnership model combines N2's action segment with N1's security (or complimentary) segment to create the efficiency–security tandem that is the basis of any collective criminal operation. Whereas a hierarchy model could assure such features by imposing a division of labor and the insulation of leaders, there is no evidence that supports this configuration exclusively. On the contrary, much more can be made and understood in regard to the communication structure of the Ciel network with a partnership and simple resource-sharing focus. In short, a boss is not needed if individuals are able to come together in a network and perform the necessary tasks to execute a criminal operation.

III. Direct and Indirect Connectivity Within the Ciel Network

Analyses of centrality in the Ciel network provide additional support for the partnership model (see Table 3.1). Centralization for the overall network

Table 3.1 Degree and betweenness centrality for the Ciel network

Node number	Degree centrality (rank)	Betweenness centrality (rank)
N1	0.417 (2)	0.591 (2)
N2	0.500 (1)	0.641 (1)
N3	0.083 (5)	0
N4	0.167 (4)	0.085 (4)
N5	0.042 (6)	0
N6	0.042 (6)	0
N7	0.167 (4)	0.023 (6)
N8	0.167 (4)	0.052 (5)
N9	0.083 (5)	0
N10	0.167 (4)	0.015 (7)
N11	0.167 (4)	0.085 (4)
N12	0.042 (6)	0
N13	0.208 (3)	0.087 (3)
N14	0.042 (6)	0
N15	0.083 (5)	0
N16	0.042 (6)	0
N17	0.083 (5)	0
N18	0.042 (6)	0
N19	0.042 (6)	0
N20	0.042 (6)	0
N21	0.042 (6)	0
N22	0.083 (5)	0
N23	0.042 (6)	0
N24	0.083 (5)	0
N25	0.042 (6)	0

was shaped more in terms of betweenness centrality (betweenness centralization = 60%) than direct connectivity (degree centralization = 42%). In both analyses, N1 and N2 were the key participants accounting for most of the centralization at the whole network level. However, whereas N1 was the central target of the law-enforcement investigation, N2 was ranked slightly higher in both degree and betweenness centrality.

How important is it to understand these centrality scores? If we were to assume the hierarchy that was argued to be in place by investigators monitoring these operations and completely ignore the communication patterns that make up the targeted network, the most-vulnerable participants in the importation ventures would be those who were most easily targeted from a traditional investigative approach. Traditional investigative approaches do not incorporate observations of aggregate relational data and, as mentioned above, the most-vulnerable participants in such an approach would likely be the drug mules. As discussed in Chapter 1, centrality may be an indication of an actor's importance in noncriminal settings, but in criminal networks, the case has been convincingly made that an actor's centrality is a sign of vulnerability since the most central actors are the most visible and, therefore, the most subjected to unwanted attention by law-enforcement monitoring and eventual prosecution—this would assume, of course, that investigators and prosecutors were observing the network from a centrality perspective.

Actor centrality should be patterned distinctly in the hierarchy and partnership models. In a hierarchy, the top organizational member (the head, the boss) would be in direct contact with the least number of other participants as possible so as to reduce visibility. The top member capitalizes on indirect access to opportunities offered by others in the organization. Delegation and subordination are the principal processes in a hierarchy and centrality would be expected to increase as we descend toward the lower levels of the organization. Thus, in a hierarchy, the top member should be low in degree centrality. In a partnership model, in contrast, centrality is shared by two or more participants who do not have a higher-level organizational member to contend with. Such participants may increase their strategic positioning in the network by also taking on brokerage positions, but their status as partners and the principal benefactors in a short-term scheme nevertheless places them in a risky position.

In the Ciel network, the assumed head of the hierarchy (N1) was too hands-on to justify his status as a leader. At the same time, the fact that the suspected lieutenant (N2) shared the same positional features within the network as N1 would in itself dismiss the premise of a neat hierarchy. The partnership model is more fitting here because it accounts for why and how key participants put themselves in a vulnerable, hands-on position while also assuring strategic control of the resources exchanged within the network at hand.

IV. Conclusion

What the Project Ciel network represents is likely to be typical of most criminal enterprise ventures—aside from the motivation to take such risks and the capacity to develop the acumen for transacting an illegal commodity or service within or beyond national borders, the level of organization found inside this network is not much more than a rudimentary example of co-offending before, during, and after the act of smuggling illegal drugs over a border. This act is the riskiest of criminal acts within the illegal drug-importation sequence—indeed, the fall of the network began with the drug seizures and arrests of the drug mules at the airport border crossings. Widening the scope, however, illustrates how vulnerability could be shared by others who may not act in the riskiest of segments, but who nevertheless expose themselves in the network patterns that are required for the complete execution of the co-offending sequence. If we turn back to past research on such matters, such basic organizational patterns appear to be sufficient for the execution of most criminal ventures. But as the next chapter will demonstrate, some forms of crime differ substantially from the typical short-term criminal enterprise venture. Accordingly, the criminal networks that emerge also take on distinct structural features.

Chapter 4
The Efficiency–Security Trade-Off

Whereas partnerships and resource sharing are fundamental features of criminal networks, the decision to co-offend comes with its pros and cons. Criminal-network participants face a consistent trade-off between organizing for efficiency or security. Most studies that explored this issue emphasized that networks emerging in covert contexts need to assure security and concealment above all. In this chapter, this claim is nuanced by highlighting a criminal network's objective and how it may influence the extent to which participants are able to invest in security.

The importance of trust and secrecy in criminal networks most clearly emerges when participants in such settings are confronted with the need to act. At some point, the hidden group must step forward and execute a crime. As discussed in Chapter 1, Erickson (1981) stressed the importance of security in secret societies. She argued that when participants in networks are obliged to choose between efficiency and security, organizational structures with a proven level of endurance and an established reputation opt for the latter.[1] Her key point was that in order to understand the structure of a network, the *conditions under which they exist* must first be appreciated (p. 189). Risky conditions generally lead participants to assure security within the network. One way the network members of Erickson's case studies achieved this was by relying primarily on pre-existing networks that formed the foundation upon which each secret society compensated for risk.

Baker and Faulkner (1993) also examined the link between risk and security within the context of price-fixing network structures. Their study revealed the importance of players operating in the peripheries of a criminal

[1]The emphasis on secret societies *under risk* is pivotal in Erickson's study in that she contests Simmel's (1908/1950) reflections on secret societies that overlooked risk, and in consequence, stressed the importance of hierarchy within such organizations. By placing risk at the forefront, Erickson establishes the variability of organizational structure in such contexts and considerably de-emphasizes the claim to a necessary hierarchy.

C. Morselli, *Inside Criminal Networks*, DOI 10.1007/978-0-387-09526-4_4,

network. These peripheral players were less targeted and less sanctioned than more central players. Remaining in the periphery was a way of protecting oneself. At the group level, having a periphery (or lacking a clear-cut core) is a way of opting for security before efficiency: reducing risk in the network does have a trade-off in that each operation and the transmission of information take longer to process across the network. Knowing that the risks associated with covert activities generally lead to the termination of the actions of targeted participants, a loss in efficiency clearly becomes an acceptable outcome for many participants. Thus, within the efficiency–security trade-off, security appears to be the predominant concern in criminal networks.

I. The Network's Objective and Time-to-Task

The efficiency–security trade-off is presented as the interplay between the need to act collectively and the need to individually assure trust and secrecy within these sensitive collaborative settings. The argument guiding this chapter is that investment in either security or efficiency cannot be assessed without considering the network's objective. The extent to which participants in a covert or criminal operation compensate for risk depends on the frequency of action required by the operation. Thus, for immediate purposes, there is less concern with the network participants' preferences for Erickson's trusting relationships and Baker and Faulkner's peripheral cushions than with how such preferences become applicable under certain conditions.

A central issue within the more general reflection on the efficiency–security trade-off is whether participants in a criminal network are capable of relying on trusting relationships or whether some have to risk dealing with uncertain others. Not all criminal networks pursue the same objective. For example, drug-trafficking or criminal-enterprise networks are typically designed for pecuniary profit. Others, such as terrorist networks, pursue ideological objectives. While many have argued that the overlap between organized crime and terrorist networks in some circumstances blur this basic distinction (Bovenkerk and Chakra 2007; Naylor 2002; Lyman and Potter 2000; Schmid 1996), clear contrasts nevertheless emerge when studying how a network's objective influences the incidence of action. Action refers to the execution of an operation that brings network participants together to realize a common outcome (e.g., the importation of an illegal drug consignment or the execution of a bombing attack). When the objective involves a monetary outcome, action in the criminal enterprise context is more limited in terms of time because participants expect a pay-off for their involvement in the network, and as a result, action must be played out within a reasonably short time frame—time is money, in short. When the objective is ideological, time is a more extensive resource and action may be prolonged—the ideological cause is prioritized over any episodic action and, as a result, network participants may lay low and wait for the right (or safer) moment to act.

The term *time-to-task* is used to refer to the interplay between time and action. Time-to-task is shorter in criminal enterprise networks than in terrorist networks. Time-to-task is longer for covert networks pursuing ideological causes because they are less often in action than are criminal enterprise networks. The structure of criminal networks reflects both the variation in time-to-task and the emphasis that is placed on the efficiency–security trade-off.

Security is an important concern for any criminal network, but not all criminal networks have equal resources to address this issue. Networks in which consistent action is a priority and time-to-task is shorter must act even when security is less than optimal. Networks that are limited in time-to-task compensate by maintaining a more efficient communication system at the core so as to assure that action takes place as quickly as possible, thereby reducing the likelihood of detection. Networks in which action may be delayed for an extended period (longer time-to-task) have less efficiency at the core, but are able to operate within more secure settings.

Shorter time-to-task, in this sense, requires a network that is shaped for greater efficiency while also assuring as much security as possible. Lengthier time-to-task shapes a network toward greater security while assuring as much efficiency as possible. Criminal networks marked by greater distance between participants (lengthier time-to-task) are more dispersed than criminal networks with shorter distance (shorter time-to-task) in which cliques and clusters of coparticipants are more easily detected and in which the removal of one member of such subgroups will likely lead to the removal of some or all of the remaining members, particularly at the core.

The analytical framework developed in this chapter was inspired by Krebs' (2001) analysis of the network responsible for the September 11, 2001, terrorist attack. Krebs maintained that *a covert network is most vulnerable when it is active* (p. 49). Networks that are *more* active are also more vulnerable and consequently less secure. Krebs assessed the terrorist network at its point of action (the events of September 11) and demonstrated how its structure centered primarily on security and how complimentary components (beyond the action group) increased efficiency.

The present argument follows that the 9/11 terrorist network was characterized by an atypical attribute that is best revealed when contrasted to criminal enterprise networks. To do so, Krebs' reconstruction of the terrorist network is re-analyzed and compared to the hashish and cocaine importation network that was targeted during Project Caviar. Binary and symmetrical matrices are used to re-analyze both Krebs' (2001) representation of the 9/11 terrorist network and the mid-1990s drug-trafficking network. Aside from centrality measures presented in Chapter 2, the analyses in this chapter also follow Kreb's approach and its application of geodesic metrics (see below).

Once again, Krebs distinguished between participants in the action and complimentary segments in the 9/11 network. The action segment represented the 19 hijackers; the complimentary segment represented 18 other individuals who were not directly involved in hijacking and crashing the airplanes, but who played key roles in transmitting information in the preparations leading up to that day. Drug-distribution networks also have their action and complimentary segments. In the Caviar network, all 110 participants were assigned roles as either traffickers or nontraffickers. Overall, 82 traffickers—those participants involved in the planning, coordination, and movement of the illegal commodity—were identified. This constituted the Caviar network's action segment. The remaining 28 participants in the Caviar network represented the "nontraffickers" or complimentary segment. These participants consisted of individuals who contributed necessary resources to the operation, but who did not have any direct involvement in the planning, coordination, and movement of the various drug consignments. These necessary resources included financial investments, logistical and equipment supply, communication brokerage, and legitimate fronts. Past research has identified individuals who supplied such resources in criminal networks under the more generic term, "facilitators" (Levi, Nelen, and Lankhorst 2004; Middleton and Levi 2004; Klerks 2001; Haller 1990).

II. Snakes and Clusters

Krebs (2001) described the shape of the action group (or hijackers) within the terrorist network as snake-like: sparse and displaying substantial distance between members (p. 46). Such a configuration illustrated a network that offered high security but a low communication flow between members. He used geodesic distances and average path lengths to demonstrate this. Krebs' average path length measure combined geodesic (shortest path) distances for each pair with the number of ways that these paths may be achieved (e.g., three ways to achieve a geodesic of four between two nodes). The final measure was an average for all nodes that was weighted by the number of options to achieve a given geodesic within the overall network (e.g., 80 ways to achieve a geodesic of two within the whole network).[2] A relatively high

[2]I thank Valdis Krebs for this clarification and Martin Everett for help in calculating this metric with Ucinet.

overall average path length of 4.75 was found for the network segment that linked the 19 hijackers.[3]

The most important aspect in Krebs' analysis of this particular metric is the change in the network's structure when complimentary participants (the nonhijackers) are added. Larger networks are generally expected to offer more space to their participants. Greater geodesic distances and average path lengths are therefore expected in larger networks. The contrary occurs in the case of this terrorist network. The geodesic range for the smaller action segment of 19 hijackers was between 1 and 9 (Krebs 2001: 46), and the range for these shortest paths diminished (from 1 to 6) after the 18 complimentary participants were added (see Krebs 2001: 48). The addition of these complimentary participants, essentially coconspirators who did not board the planes and who served as *conduits for money and also provided needed skills and knowledge* (p. 47), decreased the overall average path length by 38% (from 4.75 to 2.94).[4] This was consistent with Krebs' interpretation that the efficiency of the 19 hijackers increased with the addition of ties (or participants) beyond the action segment. The network's action segment was strong in security and weak in communication flow. The addition of complimentary participants brought "shortcuts" to the network and improved the flow of communication. It may also be assumed that the addition of the complimentary segment also placed the overall network at a greater risk of detection.

When studying the Caviar network, the transition from an action network to a complete network that incorporates complimentary participants resulted in an inverse pattern. Overall, the Caviar network included a greater number of participants than in Krebs' representation of the 9/11 network. Furthermore, the Caviar network's action segment, the traffickers, represented a majority of the network's participants. Although larger in size than the terrorist network, the Caviar network was more clustered with shorter distances separating participants. The geodesic range for Caviar's trafficking segment was smaller (between 1 and 4). Adding nontraffickers to the network increased this range, but only slightly (between 1 and 5). The overall average path length for the action segment of the network was 2.15. The addition of 28 complimentary participants, such as accountants, lawyers,

---

[3]Figure 1 in Krebs (2001) illustrates that average path length decreases if the four airplane groups (or task groups) are examined separately. This is consistent with the cellular structure that is common to terrorist networks and to assuring security therein.

[4]Rather than follow Krebs (2001) in using the addition of ties between hijackers as a second average path length measure (see p. 47), this measure was calculated for the larger "hijackers' network neighborhood" in order to examine how the addition of complimentary participants modified the hijackers' network.

legitimate importers, border agents, and other nontraffickers, increased this metric by 37%, to 2.95.

Thus, in contrast to the terrorist network, in which complimentary actors made the action segment of the network more efficient by reducing distance between participants, the addition of complimentary participants to this drug-trafficking network increased distance and therefore assured greater security for all involved.

III. Centrality Issues and Distinctions

Terrorist networks lack a core, whereas criminal enterprise networks, such as drug-trafficking operations in the Caviar case, are built outward from a core. This is demonstrated by examining variations in density and centrality from action to complete networks. Adding more nodes to a network generally makes that network less dense. Krebs reported a slight drop in density (from 16 to 13%) after the complimentary segment was added to the terrorist network. However, centrality scores produced contrary results after the complimentary segment was added to the analysis (see Table 4.1). The present analysis reproduces Krebs' assessment of centrality scores for the overall network and includes changes in centrality from the action segment to the overall network.

The terrorist network became more centralized after the addition of complimentary participants to the hijackers' segment. Degree centralization increased by 56% and betweenness centralization increased by 13%. Although complimentary participants served as shortcuts within the network, they were not central actors. The contribution of complimentary participants appears to have been indirect, namely through the enhancement of certain hijackers' positioning within the overall scheme. Indeed, both forms of actor centrality within the network were substantially transformed when the complimentary segment was added. Only one hijacker (Node 4: N. Alhamzi) retained his high status within the centrality ranks after the network expanded. Three hijackers (7: S. Alghamdi, 9: H. Alghamdi, and 10: A. Alhaznawi) became less prominent. While it was not noticeable in the action network, Nodes 2 (H. Hanjour), 13 (Z. Jarrah), 14 (M. al-Shehhi), and 15 (M. Atta) became central once the complimentary segment was added. Most of the ties that emerged after the addition of complimentary participants were located within the 13–14–15 triad, and to a lesser extent, around Node 2.[5]

---

[5] Krebs (2001) illustrates that all four of these hijackers were instrumental in coordinating meeting ties for the network's action segment (p. 46).

Table 4. 1 Degree and betweenness centrality for the 9/11 network

	Action segment		Action and complimentary segment	
Participant's name	Degree centrality (rank)	Betweenness centrality (rank)	Degree centrality (rank)	Betweenness centrality (rank)
1. M. Moqed	0.056 (5)	0	0.028 (11)	0
2. H. Hanjour	0.167 (3)	0.111 (11)	**0.278 (3)**	**0.227 (3)**
3. K. al-Midhar	0.111 (4)	0	0.111 (8)	0.011 (18)
4. N. Alhamzi	**0.333 (1)**	**0.386 (2)**	**0.278 (3)**	**0.334 (1)**
5. S. Alhamzi	0.056 (5)	0	0.083 (9)	0.007 (20)
6. A. Alnami	0.167 (3)	0	0.083 (9)	0
7. S. Alghamdi	**0.222 (2)**	0.135 (8)	0.167 (6)	**0.116 (5)**
8. A. Alghamdi	0.056 (5)	0	0.056 (10)	0.004 (22)
9. H. Alghamdi	**0.333 (1)**	**0.395 (1)**	0.167 (6)	0.081 (6)
10. A. Alhaznawi	0.167 (3)	**0.347 (3)**	0.083 (9)	0.030 (13)
11. M. Alshehri	0.111 (4)	0.131 (9)	0.056 (10)	0.012 (17)
12. F. Ahmed	0.111 (4)	0.111 (11)	0.083 (9)	0.031 (12)
13. Z. Jarrah	0.167 (3)	**0.327 (4)**	**0.278 (3)**	0.076 (8)
14. M. al-Shehhi	**0.222 (2)**	0.247 (6)	**0.389 (2)**	**0.158 (4)**
15. M. Atta	0.167 (3)	0.119 (10)	**0.417 (1)**	**0.318 (2)**
16. A. Aziz Alomari	0.167 (3)	**0.294 (5)**	0.083 (9)	0.049 (10)
17. W. Alshehri	0.167 (3)	0.209 (7)	0.111 (8)	0.080 (7)
18. W. Alshahri	0.111 (4)	0	0.056 (10)	0
19. S. al-Suqami	0.111 (4)	0	0.111 (8)	0.033 (11)
20. R. Bin al-Shibh	–	–	0.222 (4)	0.010 (19)
21. S. Bahaji	–	–	0.194 (5)	0.004 (22)
22. L. Raissi	–	–	0.139 (7)	0.015 (16)
23. Z. Essabar	–	–	0.139 (7)	0
24. A. Budiman	–	–	0.111 (8)	0
25. M. El-Motassadeq	–	–	0.111 (8)	0
26. M. Alhisawi	–	–	0.111 (8)	0.064 (9)
27. N. al-Marabh	–	–	0.111 (8)	0.026 (14)
28. R. Abdullah	–	–	0.111 (8)	0.002 (23)
29. A. Shaikh	–	–	0.083 (9)	0
30. M. Darkazanli	–	–	0.083 (9)	0
31. O. Awadallah	–	–	0.083 (9)	0
32. R. Hijazi	–	–	0.083 (9)	0.016 (15)
33. B. Alhazmi	–	–	0.056 (10)	0
34. F. Alsalmi	–	–	0.056 (10)	0
35. Z. Moussaoui	–	–	0.056 (10)	0
36. A. Khalil Alani	–	–	0.028 (11)	0
37. M. Abdi	–	–	0.028 (11)	0
Mean	0.158	0.148	0.128	0.046
Centralization	0.196	0.261	0.306	0.296

Source: Krebs (2001).

When compared with the 9/11 network, the Caviar network provided a very different image of network centralization and actor centrality. First, the addition of nontraffickers to the trafficking network resulted in increased connectivity. Without the nontraffickers, six traffickers were disconnected from other traffickers. Reachability was consistent across all participants once nontraffickers were added to the network, suggesting a first indication of how nontraffickers may insulate some traffickers. Second, centralization in the Caviar network was higher in all forms (see Table 4.2). Third, means for each centralization measure for both the action segments and the complete networks were higher in the terrorist network. Fourth, the drug-trafficking network increased primarily in betweenness centralization (by 15%) when its complimentary segment was added. This is another indication that insulation is carried into the network through the indirect relationships that nontraffickers bring to the operation.

In the Caviar network, key participants were more stable but they were also more easily detected. The addition of nontraffickers had a minimal effect on the centrality ranks of traffickers, and no new traffickers emerged as central nodes when these complimentary participants were added. Rankings in centrality scores were consistent from one measure to the next. Node 1 (the principal coordinator for hashish importations) and N12 (the principal coordinator for the cocaine importations) were the central players in the trafficking network. Two other key traffickers (N3 and N76) were less prominent in terms of centrality, but had pivotal roles in making links with various nontraffickers and in redirecting the network toward importing cocaine when hashish consignments were repeatedly seized by police.

Table 4.2 Degree and betweenness centrality for the Caviar network[a]

	Action segment		Action and complimentary segment	
Node number	Degree centrality (rank)	Betweenness centrality (rank)	Degree centrality (rank)	Betweenness centrality (rank)
1	0.531 (1)	0.557 (1)	0.550 (1)	0.637 (1)
3	0.235 (3)	0.095 (4)	0.248 (3)	0.105 (5)
12	0.333 (2)	0.342 (2)	0.257 (2)	0.292 (2)
76	0.136 (4)	0.105 (3)	0.138 (5)	0.115 (4)
87	–	–	0.147 (4)	0.119 (3)
Mean	0.039	0.016	0.034	0.015
Centralization	0.504	0.547	0.526	0.628

[a]Centrality scores are calculated with all participants included (n = 82 for the action segment and n = 110 for the overall network), but only the most central participants are displayed.

A final contrast between the terrorist network and the Caviar network concerned the centrality of complimentary participants. One nontrafficker did emerge as a central participant within the overall Caviar network—N87 was a key financial investor in various drug consignments. Aside from his central role, adding nontraffickers to the trafficking network did not introduce participants who were central in ordering the overall criminal scheme.

IV. Conclusion

The main point extending from the analyses in this chapter, and which should be remembered when reading the other case studies in this book, is that not all criminal networks have the luxury of relying on the *judicious use of transitory short-cuts* (Krebs 2001: 47). As a collective, participants in criminal networks are not generally capable of trading efficiency for security. Such flexibility is a particular characteristic of networks with lengthy time-to-task properties, as was the case with the 9/11 network. Though security is an obvious issue for all criminal networks, terrorist networks are particular in their ability to assure this condition to a maximum. Within the trade-off, efficiency is added to the secure network through the contribution of participants who operate beyond the action segment of the network.

Krebs correctly concurred with Baker and Faulkner (1993) in reaffirming that covert networks "behave" differently from normal social networks (Krebs 2001: 49), but this finding is emphasized within the atypical terrorist context. There are differences within the variety of covert networks. Criminal enterprise networks, such as the Caviar case, demand more action within shorter time frames (less time-to-task). Efficiency must be already established within the network's action segment—shortcuts are therefore not a prerogative, and distance between participants is generally restricted within the action segment of the network. The addition of complimentary participants to the criminal enterprise network at any given moment is more likely to increase the operation's security by expanding distance between participants within the network.

In terms of efficiency, the Caviar network extended from a core of closely linked participants who were primarily traffickers. An efficient flow of communication was assured within this central portion. The addition of other actors to such a core generally contributes to extending the periphery of the criminal network. Such a periphery, at least in the case of drug-trafficking networks, insulates the core. Thus, peripheral participants bring security to the network by serving as investors, masks (e.g., legitimate importers), launderers (e.g., accountants or financial advisors), possible pawns (e.g., border agents and money couriers), logistical contributors (e.g., truckers and port dockers), or brokers between disconnected traffickers.

Chapter 5
Legitimate Strengths in Criminal Networks

The findings from the previous chapter bring us to a more specific focus on how individuals from legitimate spheres may contribute to criminal enterprise networks beyond their role as peripheral insulators of the network core. Two perspectives from organized crime research help us approach this issue from opposing angles. Some perceive such individuals as pawns who are exploited by criminal-network participants. Others argue that such individuals are at the source of the problem and are far from being at the mercy of any criminal network intrusion in their legitimate activities.

The threat discourse is at the forefront of the traditional understanding of interactions between participants in criminal and legitimate forms of enterprise. Members of criminal organizations are presented as the instigating and dominating force in such relationships (see Naylor 1997; Woodiwiss 2001; or Van Duyne 2004 for elaborations on the threat discourse). This perspective was most evident in early research on organized crime, which placed considerable attention on the increasing presence of criminal trade participants in legitimate industries. Organized crime members were perceived as outsiders who forced their way into legitimate economic sectors. Their methods of extortion were too formidable for the law-abiding legitimate entrepreneur to compete. The presence and progressive domination of organized crime members threatened the values and ethics of the free market economy. Indeed, this infiltration was perceived as a greater societal threat than the apparent domination of criminal organizations in prohibited markets (Cressey 1969). More recent research has offered a similar overview of the legitimate/criminal trade overlap (see Jacobs, Friel, and Radick 1999 and Jacobs and Peters 2003 on labor racketeering in New York City).

One alternative approach, the organizing crime perspective, counters this traditional and popular view of organized crime as a predatory force within legitimate enterprise and conventional society. Rather than presenting the legitimate side of the overlap as a target of organized crime members, this counter-argument centers on the symbiotic relationships that link legitimate and criminal actors in a common setting.

C. Morselli, *Inside Criminal Networks*, DOI 10.1007/978-0-387-09526-4_5,

Chambliss (1978) found such a symbiotic process in Seattle. He exposed a crime network that structured the vices and assured impunity through alliances between criminal trade participants and a variety of legal actors: *Politicians, law-enforcement officials, professionals (especially lawyers, accountants, bankers, and realtors) and 'legitimate' businessmen became partners in the illegal industry* (p. 55). No formal criminal organization was necessary to orchestrate the system and force compliance therein; the network, in short, was maintained by the contribution of all involved.

The most explicit and complete assessment of the symbiotic relationship linking underworld activities and upperworld activities was offered by Block and Chambliss (1981) in their explanation of union corruption in labor-management relations. They contested the term "labor racketeering" because it projected an image of criminal outsiders mixing with organized labor leaders to taint labor management. This image was problematic because business actors were omitted from the racketeering framework. Block and Chambliss were also concerned with the growing trend that had *social scientists and popular writers alike [focusing] on the 'racketeers' and the 'rackets', rather than the symbiosis between business and corrupt labor practices* (1981: p. 87). They proposed the concept of *organizing crime*, which emphasizes the historical process through which organized crime arises in a given setting. Unlike traditional theoretical frameworks of organized crime that stressed the domination of criminals who orchestrate legitimate spheres of society, the organizing crime framework inverted the direction of influence by emphasizing that organized crime is rooted in the legitimate sphere of society and that legitimate actors are critical throughout the process.

Within the organizing crime framework, criminal and legitimate entrepreneurs are actors in the same political economy. Organized crime is directed by processes emerging from the legitimate spheres of society. This may be observed in one of two ways: (1) in a political economy, criminal opportunities emerge in a consistent and increasingly organized fashion to progressively transform into a sustained organized crime phenomenon that combines actors from both legitimate and criminal spheres, or (2) members of existing criminal organizations may be solicited to provide their organizing services in legitimate settings.

The first organizing crime scenario was documented in specific occupational settings, such as maritime ports (Block and Chambliss 1981; Block 1991; although not explicitly in the organizing crime framework, Mars 1983 is also a strong contribution in this general perspective). More recent illustrations that are consistent with this framework include McIllwain's (2004) study of New York City's Chinatown at the turn of the twentieth century and three studies on human smuggling channels (Zhang 2008; Zhang and Chin 2002; Kleemans and van de Bunt 2003). These studies blur the traditional image of the criminal entrepreneur as a predatory force in legitimate circles. They also maintain that the organization of crime emerges from the abundant opportunities offered by consensual legitimate actors.

The second organizing crime scenario is most evident in research following the "Mafia as a private protection industry" thesis. Rather than extortion practices, the focus here is on protection services offered by criminal entrepreneurs in symbiotic settings throughout the world. In such exchange contexts, the legitimate actor is often the initiator of the relationship and the most likely to benefit (see Gambetta 1993; Gambetta and Reuter 1995; Milhaupt and West 2000; Varese 2001; Hill 2003).

The organizing crime perspective emphasizes the consensual relationships that unite participants from a variety of legitimate economic sectors with a range of criminal trade participants. Whereas much attention has been devoted to illustrating such an overlap in either legitimate industries and economic sectors, or in practices that represent the overlap to begin with (e.g., corruption and collusion), the analyses in this chapter transpose this general outlook in an examination of the presence of actors from legitimate sectors in criminal settings.

I. Legitimate Actors in Criminal Settings

Two findings in the previous chapter illustrated how nontraffickers in the Caviar network appeared to be insulating some traffickers and how some nontraffickers were themselves central participants in the trafficking network. These complimentary participants were important in the networks because they helped shape the efficiency–security trade-off. The presence of these "legitimate" actors in criminal networks has been of some concern for criminologists studying criminal markets, organized crime, economic crime, and other forms of criminal enterprise. The overlap between legitimate and criminal settings is evident, but the presence of legitimate actors in criminal contexts is still ambiguous. Legitimate actors connected to networks designed for criminal ventures may serve as tokens or exploited prey, but they are often consensual actors pursuing their own interests. Although their place in a criminal operation is often justified by the contribution of a legitimate status or facade to the network, the personal interests that are often at stake may also lead them to have a more active and critical involvement in the criminal networks in which they participate.

Whether their presence is of a token or critical function, legitimate actors are generally perceived as *facilitators* in criminal settings. Their presence and importance have been well documented in studies on illegal drug-trafficking. Reuter and Haaga (1989), for example, described how opportunities to participate in illegal criminal ventures were widely available in legitimate work settings and how legitimate actors brought complimentary resources to criminal operations. Dorn, Murji, and South (1992) referred to such participants as "business sideliners" (see pp. 26–30) and presented them as a problematic subgroup in trafficking settings: *by virtue of their legitimate base, resources and channels and the hypothesized infrequency of their involvement, they are particularly hard to detect, contact or research* (p. 29). In such research, the contribution of legitimate actors to a trafficking network was restricted to their involvement in the circulation of a drug commodity, in which their legitimate status served as a useful front.

Actors from legitimate professions and occupational settings also bring expertise and a variety of resources to a criminal network. This was a recurrent finding in a three-city (Turin, Barcelona, and Amsterdam) survey on ecstasy markets (Gruppo Abele 2003). The Amsterdam segment of the research provided the most information on this matter. Overlaps between this illegal drug market and legitimate activities included the involvement of employees in legitimate work settings. Such personnel included not only chemists who provided laboratory equipment, chemicals, and contacts needed for drug production, but also *drivers for international trucking companies, luggage handlers, stewardesses and cleaners at the airport who can move easily on both sides of controlled areas* (p. 68). Other examples of such resource allocation were offered by Lyman and Potter (2000) who extended the diversity and importance of legitimate actors in the context of illegal drug-trafficking: *behind the scenes are many unseen workers: middlemen, financiers, smugglers, chemists, pilots, bankers, attorneys, and enforcers* (p. 223); *Indeed, without the surreptitious aid of public and private figures such as law enforcement officers, judges, prosecutors, mayors, bankers, attorneys, accountants, and elected and appointed political persons at all levels of government, the organized crime unit could not flourish* (p. 11).

The need for legitimate actors in criminal settings is well established. Much attention has centered on legal professionals such as lawyers and notaries who provide important services to criminal enterprises (Chevrier 2004; Di Nicola and Zoffi 2004; Lankhorst and Nelen 2004; Middleton and Levi 2004) and finance agents who attend to money-management and investment necessities (Duyne and Levi 2005). However, the legitimate actor is generally presented as a nominal supporter or passive service provider to the criminal entrepreneur. A closer examination suggests that we can move beyond this pawn-like image of the legitimate actor in criminal enterprise.

The term, facilitator, is generally used to represent participants in criminal activities who assist in supplying an illegal commodity or service. Such participants are not necessarily from legal or conventional settings and consist mainly of labor and management available in any "irregular economy" to supplement operational needs in criminal trades (Ruggiero 1996). Brokers in criminal trades, for example, are often described as participants who contribute to numerous criminal operations by using their social capital to facilitate the actions of suppliers and movers of prohibited commodities (Haller 1990; Klerks 2001; Kleemans and van de Bunt 2003; Morselli 2005). Aside from legitimate status, business experience, financial capital, and logistical resources, legitimate actors may also be well-positioned facilitators who offer similar brokering services in criminal networks. This is discussed by Williams (2001) who, in emphasizing the strategic positioning of legitimate actors in criminal networks, refers to them as "gatekeepers" or "crossovers." The importance of such participants suggests that it is not because a legitimate actor is taking part in a criminal venture that his role or position remains, by definition, passive to the criminal trade actor. As participants, legitimate actors may be more actively involved in how the criminal venture evolves.

Using, once again, the contextual setting offered by the drug-importation operations in the Caviar network, this chapter builds on the previous chapter and pursues this line of inquiry within the upperworld/underworld symbiosis by examining how legitimate actors facilitate ventures in criminal enterprise by focusing on their contribution to network structuring. The nontraffickers in the Caviar network represent the set of legitimate actors. At least for a minority of legitimate actors involved, the facilitating function in the criminal network went beyond the mere token role.

II. Differences Between Trafficker and Nontrafficker Subsets

A sociogram of the Caviar network is presented in Fig. 5.1. Of the 110 nodes, 51 (46%) had direct contact with only one other node in the network. The 82 traffickers had, on average, 3.9 contacts in the overall network and 6.7 contacts if we exclude the set of one-contact nodes. Nontraffickers were less directly connected with an average of 3.1 contacts in the overall network and 4.5 contacts after removing the one-contact nodes. Within the nontrafficking segment of the network, those occupying financial roles had more direct contacts than those occupying nonfinancial roles.

The Caviar network was centered on three key traffickers (N1, N3, and N12). Indeed, the most-accurate assessment of this network is that of an overlap between the networks of these three key participants. Although depicted by law-enforcement investigators as the mastermind of the overall network, N1, who was the most connected with 60 contacts, is more accurately

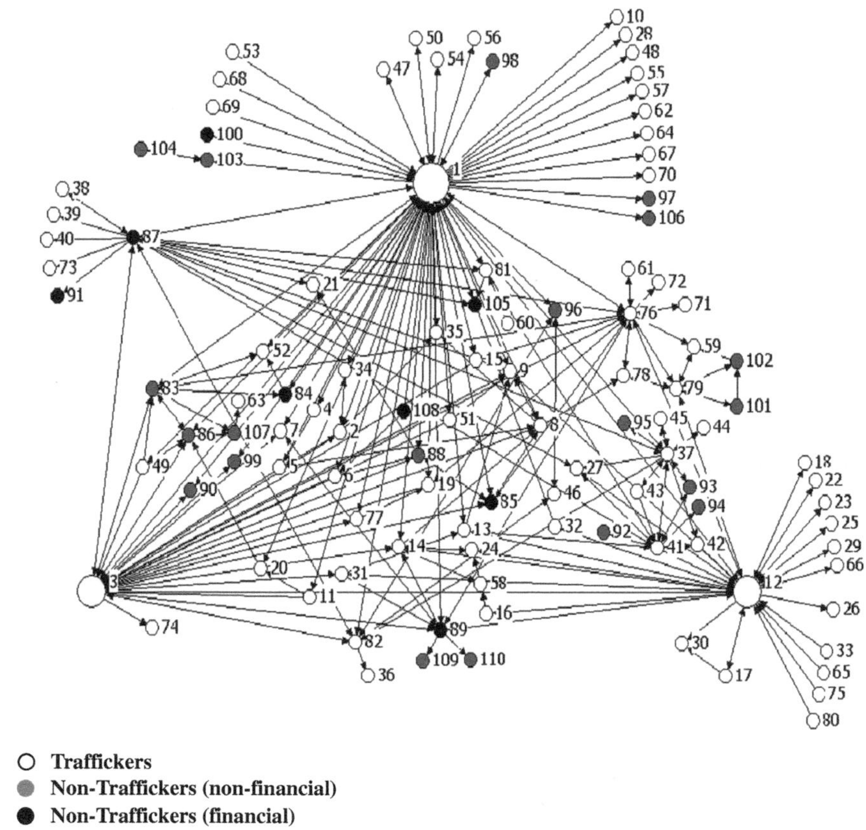

Fig. 5.1 The Caviar network

described as the principal coordinator for the hashish consignments. N12, who had 28 contacts, was the principal coordinator for the cocaine consignments. Described by law-enforcement officials as N1's "lieutenant," N3, who had 27 contacts, was extensively involved in both the hashish and cocaine consignments. Whereas 27% (n = 29) of the remaining 107 nodes in the network were not in direct contact with any of these three key traffickers, a sizeable proportion was directly connected to at least one (n = 51 or 47.7%) or two (n = 23 or 21.5%) of them. Only four (3.7%) of the participants were directly connected to all three of the key traffickers.

As a subset, the majority of nontraffickers were considerably dispersed among those nodes having either no direct contact with N1, N3, or N12 (39%, compared to 23% of traffickers) or who had direct contact with only one of the three (32%, compared to 53% of traffickers). For the remaining nontraffickers, their links to these central traffickers were slightly higher than

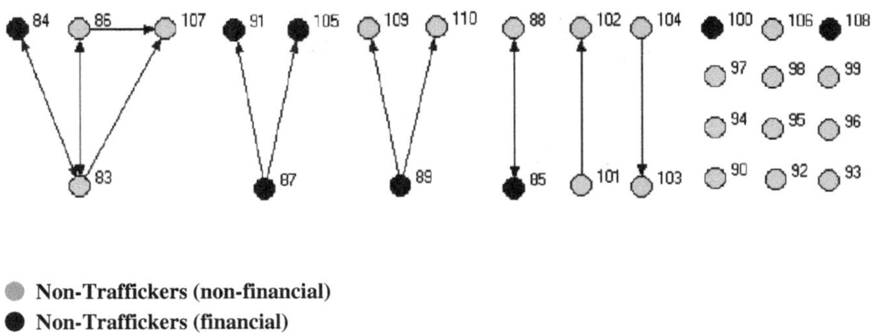

Fig. 5.2 Nontraffickers in the Caviar network

for the remaining traffickers: eight or 29%, compared to 24% in the trafficker segment, were in direct contact with two or three of the key nodes.

Thus, most nontraffickers had minimal links with the core of the Caviar network. However, this does not imply that they were not providing facilitating services for the remainder of the trafficking segment. Indeed, Fig. 5.2 shows that once all traffickers are excluded from the network, a disconnected subset of nontraffickers remains. What this illustration tells us is that the principal contacts that nontraffickers had in the network were traffickers, albeit not the three key ones.

Although not cohesive as a subset, the nontraffickers did contribute to the seamlessness of the Caviar network. Once all nontraffickers were removed from the network, the single component was divided and six traffickers (N21, 38, 39, 40, 63, 73) were isolated from the network. Their isolation was a result of the extraction of three nontraffickers who served as their only direct links to the trafficking operations: N87 was linked to five of the six isolates; N107 was linked to two; and N88 was linked to one.

These basic analyses of the Caviar network highlight important distinctions within the subset of nontraffickers. Most nontraffickers do not appear to be critical, although a minority emerge as significant participants in structuring the network. Moreover, the contribution of these select few appears to be oriented more toward the trafficking segment of the network than toward other nontraffickers.

III. Seeds in the Network

Table 5.1 illustrates how nodes were generated in the electronic surveillance data that were used to depict the Caviar network. The two principal seeds in the network were traffickers: N1, the main hashish coordinator, generated 44% of the network (n = 48); N12, the main cocaine coordinator, generated

Table 5.1 Node generation in the Caviar network

	Traffickers	Nontraffickers	Total
Number (%) of nodes	82 (74.5)	28 (25.5)	110
Number (%) of seeds	13 (61.9)	8 (38.1)	21
Number (%) of nodes generated by	88 (80.0)	22 (20.0)	110
Number (%) of traffickers generated by	67 (81.7)	15 (18.3)	82
Number (%) of nontraffickers generated by	21 (75.0)	7 (25.0)	28

12% of the network (n = 13). The third most important seed was N87, a financial investor who brought 9% (n = 10) of the nodes into the Caviar network. Nontraffickers occupying such financial roles were more important seeds than other nontraffickers. Six nontrafficking seeds were involved in the transport of money, as messengers, or as legitimate entrepreneurs and workers masking the various hashish and cocaine consignments. Only two of the nontrafficking seeds (N87 and N89) had operational roles as financial managers or investors, yet they were more important than all nonfinancial seeds in generating nodes: N87 and N89 brought ten traffickers and four nontraffickers into the network, whereas the six nontraffickers occupying nonfinancial roles brought only five traffickers and three nontraffickers into the network.

IV. The Direction of Contact

Assessing the importance of a selection of nodes in a network is also achieved by studying reciprocity (see Wasserman and Faust 1994: 510–511). A reciprocal relationship means that two actors are symmetrically linked in that they both contact each other. A nonreciprocal relationship means that two actors are asymmetrically linked in that only one initiates contact within the dyad. In asymmetric relationships, a node is either the director or receiver of contact. A relationship between a given node (A) and any of his/her direct contacts (B) may therefore be: reciprocal (A <-> B), directive (A -> B), or receptive (A <- B).

This relational categorization is applied to the Caviar network to identify the extent to which nontraffickers, compared to traffickers, were involved in reciprocal (symmetrical) or asymmetrical (directive or receptive) relationships within the network. The assumption underlying the "don't call us, we'll call you" dictum, which gains importance in criminal settings where the concealment of personal identity is fundamental, suggests that directors in asymmetric relationships are generally more important players. Participants who are more directive in their asymmetric relationships have greater control over how easily others can contact them. However, accepting this assumption means that an assessment cannot be made on whether some key players, in spite of their limited presence, can function effectively in the criminal network.

In the overall Caviar network, the reciprocity patterns of nontraffickers were similar to those of traffickers. On average, traffickers maintained reciprocal ties with 37% of their contacts compared to the nontraffickers whose reciprocal ties averaged 33%. The asymmetrical ties of traffickers represented 63%, of which 41% were directive and 59% were receptive. In comparison, the asymmetrical ties of nontraffickers represented 67%, of which 42% were directive and 58% receptive.

Nontraffickers emerged as more directive participants than traffickers when examining the most connected nodes in the Caviar network. The focus here is restricted to traffickers and nontraffickers with five or more direct contacts (n = 14 for traffickers; n = 6 for nontraffickers). Traffickers in this select group maintained an equal proportion of reciprocal and nonreciprocal relationships (an average of 50% for both). This even split is also reflected in their nonreciprocal relationships, where directive and receptive ties both average 50%. Nontraffickers, for their part, had reciprocal relationships with 55% of their contacts and nonreciprocal relationships with 45%. However, in sharp contrast to the trafficking segment, nontraffickers were more directive in nonreciprocal relationships with 72% directive and 28% receptive.

Nontraffickers who were involved in the financial aspects of the Caviar network had the greatest proportion of both nonreciprocal and directive relationships. Using reciprocity (or nonreciprocity) as a key indicator of criminal network positioning suggests that money managers and investors are as critical in orienting the network as are the main traffickers. The three most-connected traffickers (N1, N3, and N12) all had networks that were evenly split between reciprocal and nonreciprocal relationships. Yet, only N1 had a higher proportion of nonreciprocal relationships that were also directive (79%); N3 and N12 had less directive ties (46 and 43%, respectively). Of the six nontraffickers with five or more contacts, three (N85, N87, and N89) occupied financial roles. Only N85, who had the minimum five contacts, experienced more reciprocity. N89 had nine contacts, six of which were nonreciprocal, with four of these nonreciprocal relationships directed by him. N87 was the most-connected nontrafficker with 16 contacts, and just over half (n = 9) were nonreciprocal—all were directive.

It may be that the high level of directive relationships within the nontrafficker subset took place mainly with other nontraffickers; if this were so, this chapter's findings would simply reveal that such participants were directive in the legitimate circles in which they likely occupied dominant roles. This was not the case. Recall that the nontrafficking segment of the Caviar network was highly disconnected (see Fig. 5.2). If we consider the two main nontraffickers that emerged from this reciprocity analysis, N87 and N89, both were in contact with only two other nontraffickers; in all cases, these two main players were directing these relationships. This left two directed relationships for N89 and seven directed relationships for N87 that involved traffickers. Even after removing all other nontraffickers and assessing N87 and N89's position among the subset of traffickers, their relationships were still largely nonreciprocal (54% for N87 compared to 56% in the overall network; and 57% for N89 compared to 67% in the overall network). The removal of nontraffickers from the network and the slight increase in reciprocity for N89 led to some change in N89's nonreciprocal relationships: 50% of ties with traffickers were directive (down from 67% in the overall network). However, the removal of nontraffickers did not affect N87, who remained directive in all his relationships with traffickers.

V. Discrete Participants and Pawns

A participant's degree of implication and level of importance in the trafficking network can be partially ascertained by the judicial outcomes in the Caviar case. In all, 22 participants of the Caviar network were accused. Six were nontraffickers (N83, N86, N87, N88, N96, and N101). For the 14 who were found guilty and condemned to incarceration, the average prison sentence was 6.5 years. The most severe sentences were given to the principal coordinators of the hashish and cocaine consignments: N12 (15 years) and N1 (11 years).

The present analysis was designed to identify those legitimate players who brought more than their mere status as legitimate actors to the drug-importation network. The seed analysis identified the importance of the financial investor, N87, who was amongst the accused and who received one of the more severe sentences (8 years). N87 also emerged as a key player in the reciprocity analysis, as did N89 (another investor) who avoided arrest. Throughout the case, investigators centered much of their attention on an accountant, N85, who was arrested but not accused. Both N85 and N89 represented legitimate entrepreneurs who made important contributions to the operating of the criminal network (particularly in financial ways), but whose participation remained discrete. Their roles were suspected but knowledge about their degree of implication remained ambiguous throughout the investigation. These participants were identified, but they contributed little to the structuring of the network and their nondirective involvement kept their physical presence ambiguous as well.

Other legitimate players served as legitimate guises for the criminal network. N83, N86, N88, N96, and N101 all had nonfinancial roles. N83, N86, and N88 were money carriers; N96 and N101 were legitimate importers who lent traffickers their services and company name to cover up the hashish and cocaine importations. They were all under the direction of the main traffickers in the network. Their physical involvement was detected, but they did not make significant contributions. They were the pawns of the criminal network, yet they cannot be considered as naïve tokens. Their consensual implication in the minor aspects of the illegal trafficking setting exposed them to prosecution, albeit to a much lesser extent. Only two received prison sentences: N83 (3½ years) and N101 (4½ years). The remaining three were the only participants who were accused, but who were not convicted. Charges were dropped against N86 and both N88 and N96 received suspended sentences in exchange for their collaboration as informants.

VI. Conclusion

Many of the same skills and logistics are required in criminal and legitimate enterprises. Identifying an overlap between these two spheres is, therefore, not new. Much research has been devoted to highlighting the importance of legitimate workforce roles and functions in criminal operations. Law-enforcement investigators who monitored the Caviar network were fully aware of the significant involvement of legitimate actors: N85, who was the network's main accountant and a close contact to N1, was identified by officials as the principal participant emerging from a legitimate occupational setting.

The analyses in this chapter went beyond the importance of accounting, financial management, investing, document preparation, cargo movement, or a number of other tasks executed by legitimate trade players in criminal trade settings. Aside from the tasks that they execute, such participants also help structure the criminal network. Although most of the nontraffickers in the Caviar network had minor roles, a select few participated in ways that were well beyond the scope of their legitimate trades.

Identifying this select and shadowed portion of the facilitating population is crucial in determining the degree to which, as a subgroup, it contributes to sustaining criminal enterprise. The presence of legitimate actors also has its relevance for research beyond illegal drug-trafficking networks and the individual case analysis in that legitimate industries are often at the root of the key "innovations" that influence criminal opportunity structures of various kinds (for auto-theft, see Tremblay, Talon, and Hurley 2001; for cigarette smuggling, see Alain 1999). Indeed, this subset's resilience would appear to surpass that of more criminally embedded facilitators: as the criminal labor force is faced with more consistent and fervent targeting by law-enforcement officials and, consequently, a higher turnover rate, legitimate trade participants have the luxury to retract to their legal occupations and wait for the subsequent opportunity to moonlight in a criminal venture.

Chapter 6
Law-Enforcement Disruption of a Drug-Importation Network

The previous chapters have addressed each criminal network as a static object. Like social networks in general, criminal networks are not static—they are dynamic. Studying and arriving at accurate representations of the dynamics in criminal networks is probably the most challenging obstacle facing anyone approaching this area. Criminal network features may shift and change for several reasons. New criminal opportunities may emerge that require a reconfiguration of an on-going network. Competition between criminal trade participants may intensify, resulting in structural changes that indicate a network's defensive or confrontational orientation within the competition. Probably the most consistent "threat" facing any criminal network is that presented by law-enforcement controls that, at any given time, may lead to a network's stagnation, adaptation, or downfall.

In this chapter, change in a criminal network is examined with specific attention devoted to the central node issue. This is done by studying the impact that law-enforcement controls had on the Caviar network. Because the investigation that targeted this drug-importation network extended over a lengthy time frame and because data was available on how the investigation was organized, Project Caviar is the only one of the cases in this book that allows for such an analysis. By incorporating the strategies and targeting used by the law-enforcement team that made up the Caviar investigation over the two-year period, an assessment is made on how law-enforcement intensity affected the structural features and inner-workings of this particular criminal network.

I. Coding for Criminal Network Dynamics

Two sets of matrices were analyzed: the overall matrix that incorporated interactions between the 110 participants over the entire 2-year period and 11 phase matrices. These latter matrices were designed along 2-month

C. Morselli, *Inside Criminal Networks*, DOI 10.1007/978-0-387-09526-4_6,

investigative phases that matched the renewal deadlines that investigators were obliged to respect in order to continue the electronic surveillance of the drug-trafficking operations. Affidavits for each phase contained information on the identities and communication channels (telephone and pager numbers) of participants who were already part of the surveillance net and new participants who were considered important enough to be included in the surveillance net. The investigation was made up of 11 consecutive phases for which authorization was granted. These phases are used to keep track of the evolution of both law-enforcement control and trafficking operations throughout the entire investigative period.

The Context of Control

Recall (from Chapter 2) that Project Caviar was a unique investigation because it mandated investigators to seize consignments of hashish and cocaine without arresting any of the participants involved in the importations. This general and atypical context of police intervention is used as a backdrop for interpreting change in the network throughout the investigation. The extent of law-enforcement control exerted throughout the Caviar case can be assessed in three ways: the size and accumulation of seizures (a measure of imposed losses on the criminal network); the financial value of such losses; and the scope of electronic surveillance monitoring throughout the period.

In all, four consignments of hashish and seven consignments of cocaine were seized throughout the investigative period. The first seizure (300 kg of hashish) took place during phase 4 of the investigation. The most intensive period for such interventions was phase 6, during which three consignments were seized on separate occasions (two 15-kg loads and a 2-kg load of cocaine). The remaining seizures were scattered across the latter phases: a 401-kg consignment of hashish was seized during phase 7; a 9-kg load of cocaine was seized during phase 8; a 2-kg load of cocaine and a 500-kg load of hashish were seized on separate occasions during phase 9; the most valuable consignment, 2,200 kg of hashish, was seized during phase 10; and two consignments of cocaine (of 12 kg and 20 kg) were seized separately during phase 11. These importations/seizures cannot be studied independently in that a loss generally led to the creation of a subsequent importation. Furthermore, the accumulation of seizures throughout the investigative period affected the network as participants attempted to attain immediate profits while also trying to compensate for previous losses induced by the consignment seizures.

The extent of losses experienced by the Caviar network participants at the hands of the law-enforcement tandem may also be translated into monetary figures. The financial costs associated with these 11 seizures reveal the extent of damage imposed on the network. Estimating import or wholesale prices for illicit drugs is always an empirical adventure, but some consistent figures were retrievable by using price estimate data from past Royal Canadian Mounted Police reports and Desroches' (2005) interviews with high-level drug-traffickers in Canada.[1] Royal Canadian Mounted Police estimates for the wholesale price for a kilogram of hashish in Montreal during 1995 ranged from $7,000 and $10,000. The wholesale cocaine price for the same area and period was $48,000 per kg. Estimates for a kilogram of cocaine obtained during interviews with incarcerated traffickers ranged between $30,000 and $40,000 at the wholesale level (Desroches 2005: 94–95). Using midpoint values, the price for a kilogram of hashish can be established at about $8,500 and a kilogram of cocaine at about $40,000. The overall financial loss for the importation network targeted within the Caviar case is therefore estimated at approximately 32 million dollars. The breakdown across phases is telling of the extent to which network participants were consistently confronted with the challenge of making up for past and growing losses: 2.5 million dollars in phase 4; 1.3 million dollars during phase 6; 3.5 million dollars during phase 7; 360,000 dollars during phase 8; 4.3 million dollars during phase 9; 18.7 million dollars during phase 10; and 1.3 million dollars during phase 11.

A final assessment of control extends from the fact that investigators were allowed to seize drug consignments without making any arrests. Because the Caviar network participants were allowed to persist even under such challenging conditions, this offered investigators an opportunity to tap into an extensive proportion of the network. There was no way of validating whether the entire importation network fell under the surveillance of the law-enforcement tandem, but it can be confirmed that the scope of monitoring did increase from one phase to the next and that this vision was saturated by the mid-phases. Figure 6.1 includes all individuals falling into the surveillance net and displays how the level of monitoring grew from one phase to the next, increasing from 37 telephone and pager circuits intercepted (the broken line) during phase 1 to 161 circuits during phase 11. This amounted to

[1] Importation prices were not available. Wholesale prices were used instead (all price estimates are reported in Canadian dollars). While the estimates do corroborate the financial sums that appeared in the electronic transcripts for some of the importations, wholesale prices reveal an estimate of what participants in the importation network would have made if not seized. Thus, the use of wholesale prices does overestimate the importation network's financial losses. Some attempt was also made to gather data on how the losses were distributed across the ensemble of participants, but conversation excerpts relating to such affairs were too ambiguous to arrive at a clear understanding of profit allocation within the network.

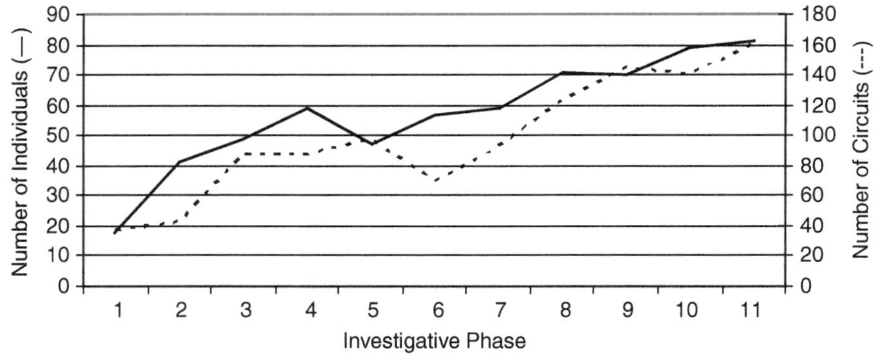

Fig. 6.1 Number of communication circuits and individuals monitored across Project Caviar phases

an increase in monitored individuals (the solid line) from 18 individuals in phase 1 to 81 participants in the final phase.

Whether in regard to circuits intercepted or individuals detected, the monitoring process follows an initial sharp rise as investigators quickly gained greater access into the workings of the importation network between phases 1 and 4. During these early phases, the volume of intercepted circuits increased by 138% (from 37 in phase 1 to 88 in phase 4) and the number of detected individuals increased even more considerably by 228% (18 in phase 1 and 59 in phase 4). Throughout the remaining phases, the scope of monitoring continued to increase, but the rise was less significant: the number of intercepted circuits increased by 64% (from 98 in phase 5 to 161 in phase 11) and the number of detected individuals increased by 72% (from 47 in phase 5 to 81 in phase 11).

As mentioned in Chapter 2, not all individuals who fell into the surveillance net were participants in the importation network. The inclusion of only the 110 individuals who were identified as participants in the Caviar network did not change the trend considerably. Figure 6.2 shows that the number of nodes across phase networks follows the same initial sharp rise between phases 1 and 4 (a 120% increase: from 15 nodes in phase 1 to 33 nodes in phase 4) and a less significant increase during the mid to later phases (a 24% increase: from 33 nodes in phase 5 to 41 nodes in phase 11).

Figures 6.1 and 6.2 suggest that while the investigators' vision of the importation network expanded at various paces until the last phase of the investigation, this outlook was gradually exhausted. I cannot establish whether this was due to the investigators' limits in breaching into the more discrete areas of the importation operations or because, after almost 2 years of monitoring, there was little of the network that was left to see.

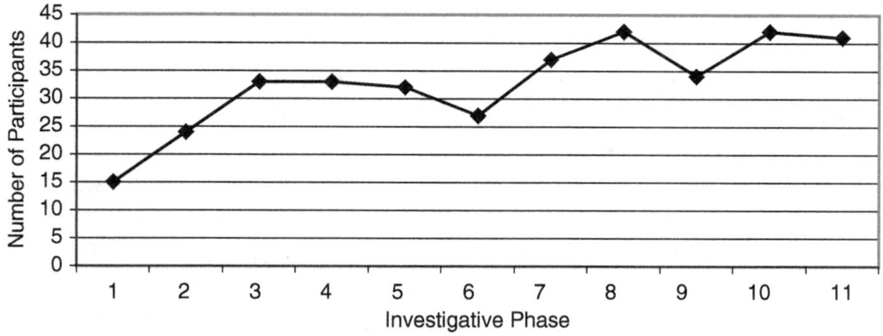

Fig. 6.2 Number of participants across Project Caviar phases

However, there is some support for the latter scenario in that in as much as the addition of circuits and individuals led to the detection of additional circuits and individuals throughout early phases, this detection process grew largely redundant throughout the latter half of the investigation as the communication patterns of new individual entries led back to already established individuals in the monitored network.

The Uniqueness of the Caviar Case

The unique quality of the Caviar case was not the network itself, but the context of control in which the network found itself. The law-enforcement officials monitoring the network that was growing and changing before them had a strategy and the resources to keep them a step ahead. For this reason, the Caviar case offers a more complete vision of a criminal network than typical law-enforcement investigations that practice the routine seize-and-arrest strategy over a much shorter time period. The seize-but-do-not-arrest strategy over an extended period enabled network expansion and the monitoring of network members as they responded to the loss of their consignments.

For research purposes, the Caviar case offered a unique experience to study how law-enforcement interventions have an impact in shaping a criminal network across time. However, that participants were permitted to (and did) proceed after such losses distorts the natural working order of the network. Few traffickers face such intensive monitoring; even fewer traffickers are permitted to continue their trafficking while under such intensive monitoring (they are usually arrested); and even fewer traffickers are confronted with continuing their trafficking operations while under intensive monitoring and while having their consignments seized. Considering the context in which the drug-importations were taking place, the case study data offers a view of a criminal network in demise.

II. Characteristics of the Overall Caviar Network

As discussed in the previous chapter, the Caviar network was a centrally coordinated network, but not to the extent maintained by law-enforcement officials who were involved in the case. Investigators repeatedly described a trafficking organization that was orchestrated by one central participant (N1), who was flanked by his main "lieutenant" (N3), and who eventually became extensively involved with another key participant (N12). The law-enforcement vision of the Caviar network was tainted by how the investigation was initiated and its lack in adjusting to change throughout the investigative period. Thus, N1 was a central participant from the start, but this was not only due to his coordinating role. Project Caviar was initiated because N1 was perceived as the individual who filled the niche left open by the conviction of the previously suspected "kingpin" who was suspected of controlling the importation and redistribution of illegal drugs in the downtown Montreal area. Thus, in the Caviar case, N1 was the first participant targeted and this initial position as a central target followed him until the end of the investigation. Yet, even though a certain level of tunnel vision was an inherent problem in the police monitoring and analysis of the network, the length and scope of the monitoring period opened the path for more observations that would not be incorporated in the final prosecution theory if the investigation ended after the first seizure. However, the network analysis of the communication transcripts do reveal shifts in actor centrality as the investigation evolved and the police interventions took their toll on the ensemble of participants.

Extending the focus beyond the involvement and connectivity of N1 revealed that this participant was indeed a principal coordinator, but primarily for hashish consignments. As with many criminal networks, the Caviar network represented an overlapping of the personal networks of a multitude of key participants. Once we acknowledge N1's over-representation in the law-enforcement assessment of the network, we see that N12 was exclusively involved in and was the principal coordinator for the cocaine consignments. Also, N3 was not N1's "lieutenant," nor was he a subordinate to either N1 or N12; he could be more accurately described as a partner to N1 and an intermediary between N1 and N12. As later sections illustrate, the Caviar network was in perpetual adjustment around these three participants.

If we rely on a static representation of the overall network, high centralization is found, as would be expected if N1 was the predominant participant. For the overall Caviar network, degree centralization was at 53%, suggesting that the network was more than a mere portrait of N1's personal network and that other participants were also occupying critical node status. Betweenness centralization was at 63%, suggesting a slightly higher level of brokerage-like connectivity.

Actor centrality scores for the overall Caviar network indicate the extent to which N1, N3, and N12 were positioned as key participants. N1 was the most central participant (degree centrality = 55; betweenness centrality = 64), followed by N3 who was equally connected (degree centrality = 25), yet less indirectly connected (betweenness centrality = 11) in his communications than N12 (degree centrality = 25; betweenness centrality = 29).

Centrality was also a key feature beyond these three key participants. Indeed, if the analysis is based simply on the final judicial outcomes of the Caviar case and the overall static representation, those participants in the Caviar network that were accused of trafficking or conspiracy offences at the end of the investigation were more central in the communication network. This is illustrated in Table 6.1: the 22 participants who were accused at the end of the investigation had higher degree and betweenness centrality scores than the remaining 88 participants who were part of the network but who did not face any criminal charges. These differences in centrality remain even after we remove the most central participants (N1, N3, and N12).

This last set of results is consistent with the observation that central actors are more vulnerable within a criminal network. Another interpretation suggests that all actors were subject to detection, but evidence for detailed criminal involvement was only obtained for the most connected. The point made here is that law-enforcement intrusion had much to do with how centrality was shaped in this network. As we will see in the next section, the various forms of centralization and actor centrality were subjected to important changes throughout the Caviar network and the static representation which has been our focus until now is insufficient to assess vulnerability.

Table 6.1 Mean differences in accusation status by degree and betweenness centrality

Accused		Degree centrality	Betweenness centrality
No (n = 88)		2.17	0.36
Yes	All Nodes (n = 22)	8.42	6.24
	Without N1	6.20	3.50
	Without N1, N3, N12	4.20	1.78
Total	All Nodes (n = 110)	3.42	1.53
	Without N1	2.95	0.96
	Without N1, N3, N12	2.53	0.61

III. Changes in the Caviar Network Across Investigative Phases

Whether through processes that are integral to the network or due to external influences, all social networks are subject to change. But the intense controls exerted on the Caviar network cannot be excluded from an assessment of change in its structure and the positioning of participants therein. Responses by participants to the drug seizures varied throughout the investigation. This is illustrated by examining shifts in tone throughout conversations between participants engaged in importations and as they became increasingly aware that they were pursuing a lost cause.

For the most part, conversations in the electronic surveillance transcripts were more representative of the daily routines and casual exchanges between participants. Only a small amount of excerpts revealed discussions surrounding the importations, but these pieces of the wiretap data do reveal explicit exchanges regarding the importation operations. The concern at this moment centers on responses to seizures. The first seizure of hashish, which took place during phase 4 of the investigation, was met with some apprehension by certain participants, but the decision to move on and attempt another importation was an immediate response. Reactions to this first obstacle were relatively calm and representative of a coping demeanor. The following conversation between N1 and N11, who was responsible for brokering the consignment from Spain, illustrates the response the day after the seizure took place (the conversation was originally in French; this excerpt is my translation):

N1: *"So we didn't make a fortune. We could always try something else from another country."*

N11: *"What are you talking about ? We lost a fortune. But I agree that we have to try something else. Did you talk with 'N12'? "*

N1: *"Yeah. First of all, stop mentioning his name. Second. I spoke with him and there's no problem. We can do something together."*

Thus, after this first hashish seizure, participants in the Caviar network re-oriented operations toward three consecutive cocaine importations—all would be seized during phase 6 of the investigation. The first of these cocaine seizures took place three months after the initial hashish bust. Reactions were, once again, calm and coping. These exchanges between N3 and N12 took place 2 days after the first cocaine seizure (N18 was one of N12's associates):

N3: *"I'm waiting for my friend to discuss the whole thing. (...) He says that if you want to try again, he would have to sit down with you. (...)"*
N12: *"Well, you know, I'd have to tell 'N18'."*

And, in another conversation later that same day, the second cocaine importation began:

N12: *"They're finding a way to relocate... not to relocate, but to redo it."*
N3: *"I wanna get down to organize with you my friend."*

This second importation, which took place about one week later, would also be seized. This time, the immediate reaction was less composed. The following conversation between N12 and N18 (one of N12's Colombian associates) occurred while both were waiting for the consignment and at the moment that N12 noticed Royal Canadian Mounted Police officials at Toronto's Pearson airport (the excerpt is a translation of the original Spanish conversation):

N12: *"The plane is landing. What's bothering me is that there are a couple of cars near the plane. They're police cars. It seems a little weird."*
N18: *"Damn it! Don't tell me that. I don't know what I'll do."*

Twenty minutes later, a customs officer made the seizure at the luggage carrousel, but N12 left before the seizure actually took place, leaving others in doubt about the whereabouts of the consignment. The following conversation took place between N12 and N18 later that same day:

N18: *"I didn't want to do it. I told you that I didn't want to do it. And you're going to tell me that everything is ok. How can you tell me again that everything is ok?"*
N12: *"What you have to do is believe me. If you don't believe and start doubting, we're screwed. (...) That idiot* (N3) *called me ten minutes ago to tell me that there was nothing there* (no police). *'You're a shit!', he told me. So I lied to him. I told him that it couldn't be—that I saw it* (the seizure). *(...) He asked me if I saw it. I lied to him. I told him that I saw it."*
N18: *"What am I going to do? Those assholes are going to think that you stole 600 lucas ($600 000). You've placed me in a horrible situation."*

The conflict surrounding the disappearance of this consignment lingered for a few more days, as this conversation between N3 and N12 illustrates:

N12: *"(. . .) Things cool off."*
N3: *"Nothing's gonna cool off. If something was done that was not supposed to be done, there's no such thing as cooling off. You cannot screw people—especially the people that you're talking to right now. These are serious people. (. . .)"*
N12: *"That's why I want to clear all this up. (. . .) I don't want to get caught that I didn't do anything about it."*
N3: *"The people down there* (in Colombia) *are jerking you around my friend."*

While relations between the various participants remained fragile, another cocaine importation was nevertheless orchestrated (and seized) later that same month. N12's realization that their operations were under surveillance was made explicit for the first time (N17 was another of N12's Colombian associates):

N17: *"What could have happened? (. . .)"*
N12: *"That idiot! You know what? It's all those calls that he made from his house!"*
N17: *"Do you really think it's that?"*
N12: *"For sure."*

During this same general period, N1 was moving back and forth between Montreal and Amsterdam coordinating another hashish consignment that would be seized during phase 7 of the investigation, about one month after the last of the cocaine seizures during phase 6. At this point, even N1 looked for ways to assign responsibility for the loss. This, however, did not deter his intentions to embark on the preparations for another importation. The following conversation took place between N1 and N69, a Dutch intermediary, four days after the seizure:

N1: *"It's their fault, because you guys did your work very, very properly. And the thing is not burnt because it could be worked again. (. . .) The guy that I know, he's a good kid. The guy in between, he's the fuck up! (. . .)"*
N69: *"It's going to harm a big company over here, which is a disaster! Because it's done by very straight people, a straight company—and there's the material, plus the damage, plus the lawyer's cost."*

N1: *"I understand. Like I told you, he's gonna make good for it. They know that because I told him—fully responsible and it has to be taken care of one way or the other because otherwise there's gonna be big problems, you know. (...)."*

N69: *"Yes"*

N1: *"Ok, listen, you want to tell me something about 'Brazil'?"*

N69: *"Yes"*

N1: *"Ok, I have the people who could work that thing out."*

N69: *"The captain, engineer... we need at least three skilled people and three unskilled people."*

N1: *"Yeah, I spoke to some people. They told me they have the skilled people. What they are asking me is how far... where do they have to go with it?"*

N69: *"I can deliver the whole thing near Dubai? And there, they go on. They don't go all the way and back. They just do the last part. I can arrange that the thing from 'Brazil' will go there, then, when it's there, you people take over."*

And so, another importation of hashish began, which would eventually be seized during phase 9 of the investigation. And before this phase, two other cocaine consignments were seized. N1's intention to make-up for the previous losses with subsequent importations was clear before the additional seizures took place, as this excerpt from a conversation with N77, another of N1's hashish importation associates, reveals: *"I'm trying to put things back on the road. I'm trying to save all this aggravation because, if it's gonna come to that, it's gonna be a mess and it's the last thing I want for you and for everybody. I just wanna fix this up diplomatically. I'm trying to put all the things together. I don't care if I don't make anything. I wanna get things started because in the future we'll do all you know. It won't be a problem."* But aggravation did spread across the network from one seizure to the next. After the seizure of the most-valuable consignment (2,200 kg of hashish) during phase 10, N1 expressed the overall predicament to N87, one of the network's financial investors: *"Now we've got to get some money somewhere because it's getting very drastic."*

Exchanges between various participants in the Caviar network not only illustrate the increasing challenges that were facing them, but also how participants were largely ignorant of the intense law-enforcement operation that was targeting them. Designation of blame was generally directed within the network and the systematic response was to begin preparation of another importation venture.

Decentralization and Core Changes

Within this context of intensive controls and increasing aggravation, the Caviar network gradually decentralized (see Fig. 6.3). During the early phases of the investigation, both degree and betweenness centralization hovered, respectively, around 80 and 90%. By phase 4, the first seizure had taken place and decentralization set in. Degree centralization dropped to 44% during phase 8 and to under 30% for the remainder of the investigation. Betweenness centralization remained around 50% after a considerable drop during phase 6 and decreased to a low of 33% during phase 10.

Centralization is sensitive to network size and the decentralization process that is observed may be a mere artifact of the rise in the number of individuals detected as the investigators accessed an increasingly wider representation of the importation operations (recall Figs. 6.1 and 6.2). But, as mentioned earlier, the increase in network size was less significant after phase 4. The drops in both forms of centralization between phases 5 and 11 were due to coordinating shifts within the network itself and the main reason for this drop was the introduction of an alternative distribution chain that imported a new product (cocaine) and that required the resources of a different set of participants.

Changes across phases and as seizures began to take their toll also emerge when studying the evolution of centrality of key participants. Figure 6.4 presents degree and betweenness centrality scores for N1, N3, and N12 across the 11 investigative phases. The illustration is telling the rise of other participants as of N1 lost influence as a coordinator.

N1's degree centrality gradually dropped between phases 4 and 7 and decreased significantly for the latter four phases. N1's betweenness

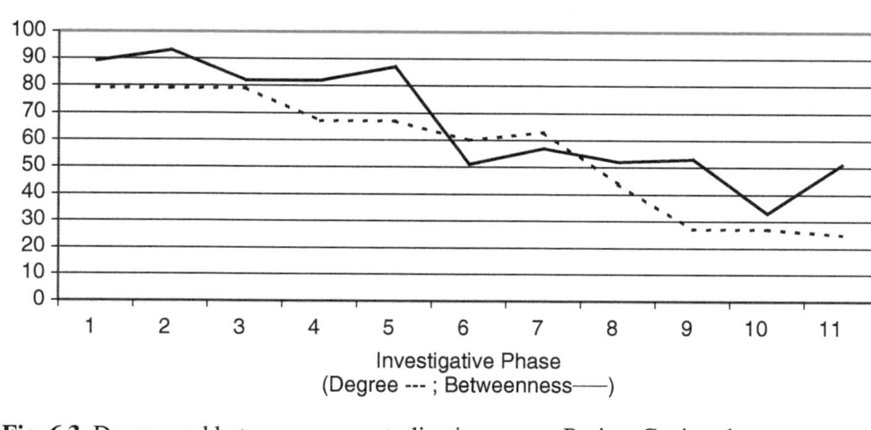

Fig. 6.3 Degree and betweenness centralization across Project Caviar phases

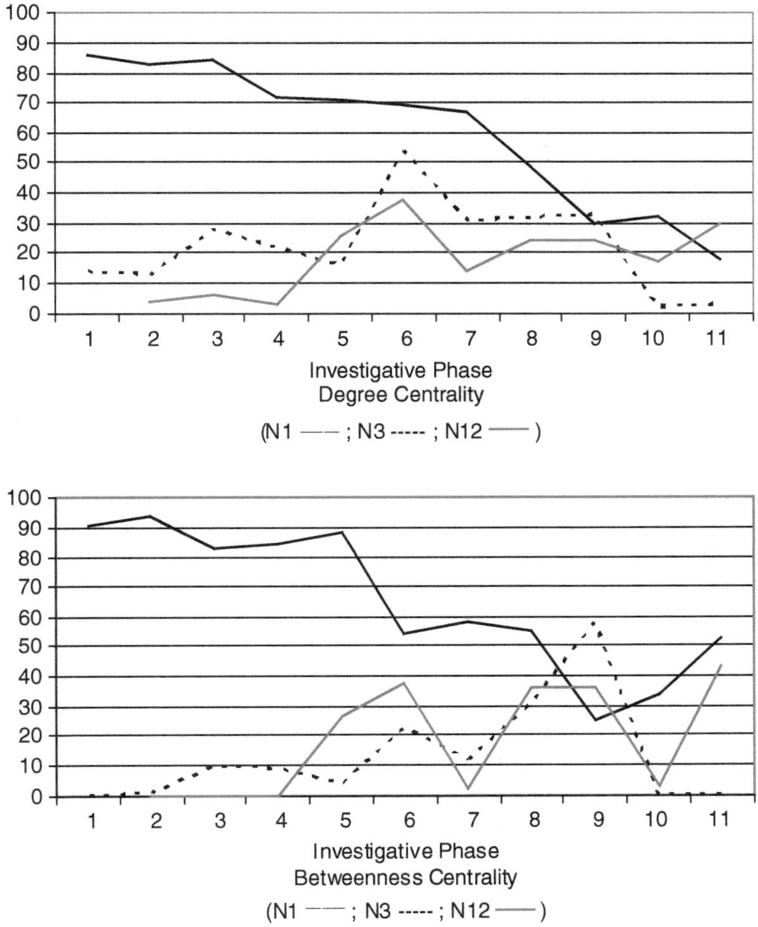

Fig. 6.4 Degree and betweenness centrality for central participants across Project Caviar phases

centrality was somewhat more erratic, but the decrease was nevertheless clear: after a sharp drop between phases 5 and 6, his position as a key broker remained well under 60%, with a low of 25% during phase 9.

The decrease in N1's central positioning coincided with increases in N3 and N12's degree and betweenness centrality scores. Both N3 and N12 appeared as marginal participants throughout the earlier phases, but, by the mid phases of the investigation, their positions as central participants within the network approached that of N1. Each participant experienced important increases in both forms of centrality during phase 6. During phase 9, N3 and N12 surpassed N1 in betweenness centrality and maintained similar degree centrality scores as they attempted to recuperate for past losses by orchestrating two cocaine importations without N1 (but N1 did coordinate a hashish importation with the participation of N3 during this same period). N3's centrality status was reduced to the periphery of the network after phase 9, partly due to the fact that both cocaine consignments were also seized and that he was largely excluded from the importations that took place during the last two phases. During these final phases, N1 would attempt the most sizeable hashish importation during phase 10 and, with N12, would attempt two more cocaine importations during phase 11.

While it cannot be confirmed that there was a clear competition for central positioning during these latter phases, it can be established that the drop in the key participant's (N1) centrality during the early and mid phases and the rise in the centrality of others were due to previous and accumulating disruptions within the network. N3 and particularly N12 gained in prominence as it became increasingly evident that importations coordinated by N1 systematically failed.

Disorder and Accountability

This demise brought N1's status within the network under assault by a much greater number of participants. Responses to drug seizures within the network were not only observed by shifts in central positioning between key participants. The communication patterns surrounding N1 across the investigative period also suggest that his prominent positioning within the network diminished at a more general level. Table 6.2 presents the number of calls made to-and-from N1 for each phase. As a key coordinator, we would assume that he would be in control of the communication network surrounding him. Being in control would imply that he was able to reach as many people as possible, while keeping the number of people that were able to reach him to a minimum. The overall number of outgoing calls from a key coordinator in control of his position within the network should therefore be higher than the number of incoming calls.

For most of the phases, N1 made more calls than he received within the communication network. Phases 6 and 7 were the exceptions with N1 receiving about twice as many calls than he initiated. The number of calls to N1 during these two phases was also unmatched during any other phase. This increase in the intensity of calls may be due to a few participants who were extremely active in contacting N1, but this was not the case. The increases

Table 6.2 Communications to-and-from N1 across phases

Phase	Calls to N1	Calls from N1
1	6	29
2	17	92
3	26	191
4	73	256
5	45	80
6	316	137
7	191	100
8	21	162
9	21	74
10	40	78
11	15	22
Total	771	1,221

during phases 6 and 7 were due to a greater number of participants who initiated contact with N1. Whereas only a minority of participants was able to initiate contact with him between phases 1 and 5, 63% of participants in phase 6 and 60% of participants in phase 7 initiated contact with him. As with the number of calls directed to-and-from this key node, attention surrounding him decreased throughout the latter phases as other participants became more central. The greater proportion of participants making calls to N1 accounts for why his degree centrality remained stable during phases 6 and 7 while his betweenness centrality dropped substantially (recall Fig. 6.4)—N1 continued to be directly connected with an extensive proportion of the Caviar network, but he became less efficient in this connectivity, resulting in a decrease in his value as a broker.

Thus, N1 was still heavily connected during this period, but the quality of his connections had taken an important shift. After five seizures in as many months, N1's status as a coordinator was increasingly contested, not only by other central participants, but by most participants within the network. Whereas during the early phases of the investigation law-enforcement officials had tapped into a network that was well ordered and organized to execute a hashish importation, their intrusion forced the network to execute subsequent importations in a less orderly manner. The consistent seizures by law-enforcement officials had weakened the central node's status within the importation network. By this point, not only could the main coordinator reach most participants, but accountability for past losses led to most participants reaching out toward him.

IV. Conclusion

Because the main concern of research on vulnerabilities in criminal networks was to promote the value of social network methods and concepts for a practical strategic analytical framework, researchers overlooked the fact that law-enforcement controls are integral in shaping criminal networks. Vulnerability comes in many forms and the main threat facing criminal networks is the targeting that such networks confront from law-enforcement agencies. The analyses in this case study guide us in understanding how drug-trafficking networks respond to law-enforcement disruption. They may prove helpful for researchers interested in the dynamics of criminal networks as they may prove useful for those seeking to target criminal networks and who follow Dorn's (2000) sound grasp that when targeted, *it is necessary to aim above and beyond it* (p. 304).

The Caviar network is an extreme case in point for analyses of law-enforcement disruption in criminal networks. The first finding illustrated that multiple coordinators took part in the network and that centrality was a key feature of individuals occupying such positions and a key factor in accounting for eventual accusations by court authorities. But centrality within the network was subject to change. Thus, the second finding confirmed past research on social organization in criminal markets and revealed that the criminal network decentralized as law-enforcement monitoring and drug seizures took their toll. This decentralization process is linked to the third finding that revealed the emergence of new central nodes at the expense of the initial central node. Such shifts cannot be disassociated from the intensification and prolongation of controls on the network. The third finding illustrated how a criminal network becomes less orderly when intensely controlled. The emergence of an environment of discontent and mistrust subjected the network to not only greater risk on the part of desperate participants looking to compensate for previous losses, but also to a greater rise in potential informants who saw the benefits in cooperating with law-enforcement officials to disassociate themselves from a series of ventures that were ill-fated to begin with (two participants became informants at the end of the investigations).

Throughout the chapter, the interpretation of change in the Caviar investigation offers a bleak outlook of the capacities of drug trade participants to react systematically to intense targeting, but an alternative view may also be formulated. After all, the network participants did prove their resilience by remaining active for almost 15 months after the first hashish seizure. Furthermore, the accumulation of additional seizures did not appear to deter the preparation and execution of subsequent importations. Even the portrayal of the main participant's positioning throughout the network was presented negatively. It may well be that N1 never "lost" his central (and vulnerable) position and that he chose to step back within the falling network and let others take on more central positions. Clearly, the initial central node's status within the network was weakened, but neither the network patterns nor the transcript contents help us confirm whether this was forced upon him or strategic on his part.

Deciding to emphasize the negative outlook was based on the context of control. Whether the central participant was or was not an active decision-maker in his loss in centrality is largely irrelevant in light of the monitoring that followed the entire network. The weakening in his centrality was due to the network's disruption by external forces. What is most striking is that although this case study's findings center on the centrality issue, no police strategy during the 2-year investigation was implemented to arrest (or remove) any of the central nodes. This only took place at the very end of the investigation. Instead, manipulating the criminal network involved the removal of the common item uniting all participants—drug consignments. Once the consignments were controlled, network participants took it upon themselves to deal with central node issues.

Chapter 7
Brokerage Qualifications in Ringing Scripts

With the exclusive focus on network structure and individual positioning, an important feature that has been largely overlooked across the case studies in the previous chapters concerns the variety of tasks and roles that participants have in crime-commission processes. Indeed, some roles may be more indispensable than others in the execution of a given crime. Such task assignment and management become a greater challenge in the context of more complex criminal ventures that require the execution of a sequence of distinct crimes. In the case of the Caviar network, for example, the sequence for the illegal drug-importation operations included the exportation, importation, and, if seizures had not taken place, distribution of either the hashish or cocaine consignments. Surrounding this main channel of the crime-commission process was a variety of other roles that combined the efforts of financial investors, money carriers, finance managers, and money launderers. In such complex ventures, interactions between participants occupying distinct roles cannot be fully understood without considering the position of network participants who coordinate and sustain the overall sequence of the crime-commission process.

In this chapter, the coordination of such processes is examined by merging crime script and social network analyses. This merge allows us to reconfigure criminal ventures (the resale of stolen vehicles or *ringing networks*, in this case study) by adding role allocation to the network framework. Within this framework, a common bond is found in the brokerage positions that are occupied by participants who are crucial for enhancing flexibility within the overall crime script. The removal of key brokers, in turn, would severely diminish a script's flexibility, thus reducing coordination opportunities for all other participants.

C. Morselli, *Inside Criminal Networks*, DOI 10.1007/978-0-387-09526-4_7,

I. Crime Scripts and Flexibility

All crimes may be drafted for their step-by-step procedural make-up. Cornish (1994) introduced script analysis to criminologists as a framework for enhancing our understanding of general crime-commission processes. The approach extends from previous theories in cognitive psychology and earlier work on specific crime modeling for situational crime prevention and a general rational choice perspective (Clarke and Cornish 1985). The script approach emphasizes the procedural nature of criminal behavior at the individual or group level: *the fact that the activity is goal-oriented; that it consists of a sequence of steps or sub-goals; that the separate elements in the sequence form themselves into a procedure which can be carried out without much thought; and that the activity requires both knowledge and the experience gained by practice for its successful performance* (Cornish 1994: 157).

Crime script analysis is primarily a tool for *generating, organizing, and systematizing knowledge about the procedural aspects and procedural requirements of crime commission* (p. 160). It is useful for researchers who are interested in studying criminal behavioral patterns within an "event schema"—such a schema would be comprised of an offender or group of co-offenders pursuing a series of subgoals to achieve a wider criminal objective. The script is a map of a given criminal procedure and is made up of *scenes* that represent the episodes or "logistical steps" which occur along the overall schema. For example, Cornish (1994) identified five scenes that comprise the sequential order of a ringing script—cars have to be: (1) stolen, (2) concealed, (3) disguised, (4) marketed, and (5) disposed. Each of these scenes may be broken down into a variety of *facets* that consist of different ways in which a scene may be executed. The theft scene may take place within a parking lot or through rental fraud. The conceal scene may direct the stolen vehicles into lock-up garages or car shops. The disguise scene may involve a change in license plates or more elaborate document switches using the identities of crashed vehicles. The market scene may incorporate a number of facets that prepare the stolen vehicles for the ultimate sale. The disposal scene executes this final sale to local buyers or buyers in foreign countries.

The ensemble of combinations between facets and across scenes is referred to as *permutation* within the script. The level of permutation represents the degree of flexibility in the criminal procedure. A high degree of permutation (or flexibility) is found in a script that provides *a number of crime-commission routes to the same outcome* (Cornish 1994: 173). Within the script, a scene that offers more facets offers more options to its participants and will therefore increase flexibility within that segment of the script. Thus, scripts with scenes that have numerous facets offer greater flexibility than scripts made up of scenes with only single facets. In technical terms, a

1 × 2 × 1 × 2 script has four scenes with minimal (one or two) facets and a low maximum number of combinations to realize a criminal objective (the product for all facets in all scenes is four). Such a script is low in flexibility. More flexibility is offered in a 4 × 4 × 5 × 4 script that also has four scenes but a greater variety of facets within each and a higher maximum number of combinations to realize the criminal objective (the product for script permutation in this script is 320).

Ekblom and Tilley (2000) advise us not to approach crime scripts as linear processes ("like books"), but as multi-optional scene sequences that offer flexibility to those involved ("like computer games"). High resourcefulness is an indication of more effective offending and this is particularly important for understanding how offenders become organized in more sophisticated social configurations. Whether in terms of resourcefulness or flexibility, Cornish offers a valuable heuristic device to assess such issues in various ways. But whereas Cornish's promotion of script analysis stressed its utility for examining the "cognitive structures" underlying action sequences and for *thinking about how knowledge of crime commission might be organized and used by offenders* (1994: 170–171), the present interest in script analysis concerns another set of choice-structuring properties that Cornish grouped under a general guise of "casting requirements."

II. Merging Crime Script and Social Network Frameworks

Casting in a script entails the matching of participants to actions required in the execution of each scene. In complex or group-level crimes, the structure of coparticipation across a crime script is intrinsic to the level of permutation because the number of crime-commission routes leading to the collective outcome depends on some participants who are able to pivot between various facets from one scene to the next. The presence of such participants influences a script's flexibility. Addressing such a proposition requires a greater focus on the network features that are at play in such crime-commission processes. Cornish did concede that networks were important for crime scripts, but, once again, his emphasis was more on how the "structure of knowledge" was an independent aspect to the people and relationships maintaining the script (1994: 185). However, the network is a fundamental feature in crime scripts and placing a greater focus on it will help identify intervention points across various crime-commission processes.

Flexibility is key in crime scripts and criminal operations that are high in this feature offer a more adaptable and resilient action setting for participants. Increasing the facets within a script's scenes makes that script more flexible. Previous research has addressed the flexibility (or adaptability) issue in crime settings, but none has benefited from a framework that unites crime script and social network analysis. Lacoste and Tremblay (2003), for example, argued that the strength of check fraud groups depended on how coparticipants interacted in coordinating script changes, which they interpreted as innovations. These check fraud script innovators subsequently served as models for a wide range of offenders involved in the same activity. They found that groups that changed various elements in their scripts experienced higher gains, lower costs, and were more geographically mobile in their past experiences and personal network extensions.

In another study, Tremblay, Talon, and Hurley (2001) examined stolen-vehicle resale (or ringing) groups in Quebec. Their study revealed the participation and importance of auto-industry professionals and legitimate scrap-yard dealers. Because of their presence, the availability of crashed vehicles became a common fixture in such scripts. Such availability increased the likelihood of body switching methods (or changes in vehicle identification) in the script process. Most importantly for our present objective, they maintained that the presence of such "moonlighters" from the legitimate spheres of the auto-industry meant that the principal participants in ringing networks were not likely located at the extreme scenes of the ringing script (theft or disposal scenes). While it was unclear what they meant by principal participants, Tremblay, Talon, and Hurley's justification appears to be that such legitimate trade representatives became so important in ringing networks that the crime-commission process could no longer be completed without them. Their importance in the script was directly related to their position as discrete yet necessary participants in the network that was active in that script.

This matter of network positioning was studied more closely by Bruinsma and Bernasco (2004), who placed a more specific focus on the organization of networks designed for the trading of stolen cars in the Netherlands. Like Tremblay, Talon, and Hurley (2001), they emphasized the service-orientation of such networks for meeting the demands of a domestic or foreign public market. Such demand generates a systematic chain-like structure that links the various "clusters" of stolen car trading. Bruinsma and Bernasco identified three such clusters: (1) thieves who stole cars; (2) a group of participants who recycled cars and prepared the necessary documents; and (3) a final group who represented demand and who organized couriers responsible for bringing cars to their final destination. Although Bruinsma and Bernasco did not enter into the subtleties of a script analysis, their analysis did offer some observations on the structure of action within and between links in the distribution chain. They found that contacts between links in the chains embracing stolen-vehicle trading were instrumental, low in frequency, and not tightly knit. Within links, contacts were more tightly intertwined, although not representative of affective relationships (p. 89). Lacking a central cluster and represented by a chain-like distribution network of somewhat cohesive links that were instrumentally attached to other links, we would expect some participants to contribute significantly to keeping these consecutive scenes attached. Bruinsma and Bernasco maintained that the "lynch pin" of the chain was to be found amongst some participants in the recycling cluster that coordinated supply and demand, assured financial settlements, and generally kept the chain network flowing. These lynch-pin participants were the brokers within the chain.

The network-script merge offers a complete framework for studying permutation and flexibility along a crime-commission process. Just as *the script offers a useful analytic tool for looking at behavioral routines in the service of rational, purposive, goal-oriented action* (Cornish 1994: 159), merging the script and social network framework offers a key analytical hybrid for identifying specific positioning within the scope of behaviors inherent throughout the crime-commission process. This is examined in the context of two ringing operations that were coordinated from Montreal, Canada, during the late 1990s.

III. The Case Study Design

In this case study, script and network frameworks are merged to help understand the broker impact in the crime-commission process. Two working hypotheses are pursued. The first maintains that if a small set of participants with high brokerage positioning are removed from chain-like criminal operations, flexibility in such operations is likely to be reduced. The second hypothesis substantiates the first. It maintains that the removal of a large set of nonbrokers should not have an impact on the level of flexibility in the criminal operations at hand.

The data sources used for this case study were derived from Projects Siren and Togo, two investigations that targeted stolen-vehicle exportation operations that were underway in the Montreal region during the late 1990s. The Siren network was made up of 44 participants. The Togo network was made up of 33 participants (see Chapter 2 for more details). Once network participants were identified and script roles were assigned, the network was merged with the script process.

Determining Brokerage Qualifications

Betweenness centrality and Gould and Fernandez's (1989) brokerage leverage were used for assessing brokerage positioning within each network. Gould and Fernandez developed their measure as an alternative to betweenness centrality. The authors argued that betweenness centrality was too reliant on geodesics within the network. In large networks, node positioning along geodesics are less relevant since these shortest paths are generally longer than geodesics in smaller networks and the value of brokerage resources is diminished by the distance separating nodes at either end of a path. In short, serving as a broker in a short path is more advantageous than serving as a broker in a long path.

Gould and Fernandez's brokerage leverage measure accounts for brokerage between subgroups within a network rather than brokerage between actors in a nonsegmented network. Subgroups in a network could be indicated by actors' affiliations, occupational roles, social class status, and a number of other categorizations. In this case study, the scenes in the ringing scripts were used to create subgroups. This permitted an analysis of how some participants were crucial for mediating the multiple channels across a script.

Another contribution of Gould and Fernandez's measure was its construction of a five-item brokerage typology that distinguished how brokers may be positioned within and between subgroups. *Coordinators* brokered within subgroups. *Consultants* were outsiders who brokered between members of the same subgroup. *Gatekeepers* and *representatives* brokered between an outsider and insider of his/her subgroup. *Liaisons* brokered between members of different subgroups. Gould and Fernandez developed two measures of total brokerage leverage within this typology. Total *raw* brokerage leverage, or the sum of brokerage leverages for all specific brokerage types, was a straightforward measure of how nodes were positioned to mediate between subgroups in a network. Total *relative* brokerage leverage adjusted for network size and subgroup size to compensate for the fact that nodes who maintained roles in the larger subgroups would have a greater chance to broker. The relative brokerage score standardizes for size by dividing the raw brokerage scores by the expected brokerage scores (that which would be predicted under a chance model). Because the main concern of this case study was to identify the key brokers in both networks, I verified whether important differences in brokerage score rankings were found when using either measure. At the top level of brokerage scores, raw and relative brokerage measures identified the same individuals. I therefore opted for the straightforward total or raw brokerage leverage. The scores presented in subsequent analyses indicate the total number of times a node brokers in any of the specific ways identified in the typology. Thus, if a node was found to broker once as a coordinator, once as a gatekeeper, and twice as a liaison, he would have a total raw brokerage leverage score of 4.

Strong correlations were found between degree centrality and the two brokerage measures for each of the cases. In the Siren case, degree centrality was highly correlated with both betweenness centrality ($r = 0.84$) and total brokerage leverage ($r = 0.83$). The overlap was even more notable in the Togo network with correlations of $r = 0.98$ between degree and betweenness centrality and $r = 0.94$ between degree centrality and total brokerage leverage. Strong correlations were also found between Freeman's betweenness centrality scores and Gould and Fernandez's total brokerage leverage. For the Siren network, the betweenness centrality distribution varied from 0 to 80%[1] and the brokerage leverage distribution varied from 0 to 994. A correlation of 0.96 was found between both measures. For the Togo network, the betweenness centrality distribution varied from 0 to 78%[2] and the brokerage leverage distribution varied from 0 to 322. The two brokerage measures were correlated at $r = 0.97$. Although there is full indication that the two brokerage indicators are, in large part, measuring the same patterns, a subsequent analysis of the impact of key brokers does reveal that the minimal difference is meaningful enough to warrant the application of both betweenness centrality and total brokerage leverage.

Assessing Participant Removal Impact on Script Permutation

This case study was designed to assess the impact of brokerage in ringing operations. The final set of analyses examines how the maximum number of combinations in script permutation, and thus flexibility, may be influenced by the removal of participants with various degrees of brokerage in each network. The aim, here, was not simply to disrupt the network by breaking it into fragments or smaller components, but also to disrupt the crime-commission process that was executed during these ringing operations. Disrupting both the network and the crime-commission process required a strategy that began with the removal of key participants and resulted in the loss of facets in the crime script. A facet of a scene was considered lost when there was no longer any connection (or any direct communicating link between participants) to any other facets throughout the script's principal component. Thus, all isolated facets were excluded in the script's permutation as were, in some cases, multiple (two or three) facets that were only connected to each other. A scene was considered lost when all of its facets

[1] Betweenness centralization in the Siren network was 79%, suggesting a working structure that revolved considerably around one or two individuals who managed indirect connectivity therein.

[2] Betweenness centralization in the Togo network was 76%, suggesting a working structure that was also heavily reliant on one or two brokers.

(and participants) became disconnected. At this point, the crime script was considered broken.

IV. Criminal Network Flexibility and Script Permutation

Some participants within and across scenes in the ringing scripts were connected to other participants in ways that offered alternative routes for moving stolen vehicles across the transaction network. Those participants with greater brokerage capital contribute greater flexibility to the script. Brokers in the network may therefore be understood as key articulators within the social network that embeds a crime-commission process and the removal of such participants should have a detrimental impact on the overall operation.

Overall Scripts

Both the Siren and Togo operations were consistent with Cornish's (1994) breakdown of ringing scripts (p. 174). The ringing script began with the theft of vehicles that were subsequently concealed, disguised, marketed, and disposed (see Table 7.1).[3] Both scripts contained a similar number of facets within each of these scenes, although the Siren script did benefit from a greater number of disposal choices with six foreign buyers and one domestic buyer (compared to three in the Togo operation). Largely because of this greater number of disposal channels, the Siren script ($3 \times 2 \times 2 \times 3 \times 7$, for a maximum of 252 combinations) was assessed as more flexible than the Togo script ($3 \times 2 \times 2 \times 2 \times 3$, for a maximum of 72 combinations). In this section, the diverse facets in both scripts are presented. Estimates for the financial costs associated with each scene are also provided through data that was retrieved from interrogations of some of the participants and from the content of various telephone exchanges that were intercepted during the two investigations. Overall, the cost for exporting a stolen vehicle hovered at about $15,000. The breakdown is presented below.

Within scenes, facets were comparable. Thefts took place in either airport or sugar camp parking lots. Rental fraud was also an important theft facet in both scripts. The approximate cost for the payment of thieves per car was $2,500. An additional cost of $750 per month was allotted to owners of

---

[3] Another scene, financial investment, was also identified, but was excluded because it did not fit within the consecutive order of the script—investors could appear at any time. In terms of network content, the loss is minimal with only one individual in the Siren network and two individuals in the Togo network acting as investors.

Table 7.1 The ringing scripts

	(1) Theft/fraud	(2) Conceal	(3) Disguise (body-switch)	(4) Market	(5) Disposal
Siren	(a) Rental fraud	(a) Warehouse	(a) Amateurs	(a) Completion of vehicle registration forms	(a) Russia
	(b) Airport parking lot theft	(b) Private residences	(b) Auto industry professionals	(b) Vehicle dispatch	(b) Egypt
	(c) Concession fraud			(c) Acquisition of blank vehicle registration forms	(c) Toronto
					(d) Iraq (e) Italy (f) Ghana (g) Switzerland
Togo	(a) Airport parking lot theft	(a) Warehouses	(a) Auto industry professionals	(a) Vehicle dispatch	(a) France
	(b) Sugar camp parking lots theft	(b) Private residences	(b) Amateurs	(b) Acquisition and completion of vehicle registration forms	(b) Ghana
	(c) Rental fraud				(c) southern Quebec

warehouses and private residences that were used for concealing the stolen vehicles in the second scene.

The combination of disguise and market scenes cost about $5,500 per car. In both the Siren and Togo cases, the disguise scene in the ringing script consisted of modifying the vehicle identification number (VIN) for each vehicle—this is the step that Tremblay, Talon, and Hurley (2001) referred to as "body-switching." Whereas the switch in VIN was a simple task in itself, getting to that identification number created problems. Access to the VIN requires the complete removal of the vehicle's windshield. In the Siren case, this task was initially executed by participants who lacked experience (amateurs) and repeatedly broke the windshields. After facing costs due to repeated windshield breaks, a professional in windshield replacement joined the network (see Fig. 7.1). This shift from amateur to professional body switchers is consistent with the "innovative" feature of ringing networks that was the center of Tremblay, Talon, and Hurley's (2001) study. The Togo operation integrated professionals in the disguise scene from the start and to a much greater extent (six participants from the auto industry participated in this scene; see Fig. 7.2). In addition, those amateurs in the Togo case who also executed the body-switching task appeared to be more competent than the amateurs in the Siren operation.

The market scene included all facets that dealt with the preparation of the vehicle for its ultimate movement and sale across an international border or within the domestic setting. In both cases, key players were identified in this scene. To a great extent, the facets in the market scene combine to represent a sub-script. For example, in the Siren operation, one participant was responsible for retrieving blank vehicle registration forms from Quebec's vehicle registration bureau (facet 4c in Table 7.1). These forms were subsequently filled by another group (facet 4a) who prepared the false registration and vehicle history certificates. Participants in the third facet (4b) took care of dispatching the vehicles to foreign and local buyers. The main coordinator (Node 1 in the Siren network; see Fig. 7.1) was located in this segment of the script. The breakdown for the market scene was similar in the Togo operation. The only difference was that the acquisition and completion of vehicle registration and history forms were executed by the same participants—this overlap constituted the only example of multiple-tasking in both cases. As in the Siren operation, the main coordinator was also found within the dispatch facet in the Togo case (Node 1; see Fig. 7.2).

The final disposal scene involved local (Toronto and Southern Quebec) and international channels for the sale of stolen vehicles. International shipments all left from the Port of Montreal. With seven destinations, the Siren script was more adequately set up. Dispatching vehicles to foreign countries by sea cost between $5,000 and $10,000 per vehicle, depending on the destination. This cost was at times reduced by filling a transport container with more than one vehicle (a container could hold up to three regular-sized cars).

Networks Within the Scripts

Once the scripts in each ringing operation were reconstructed, the second step was to transpose the two networks of participants that were acting out the various facets across scenes (Figs. 7.1 and 7.2). Recall that the Siren network contained 44 participants, compared to 33 participants in the Togo network. In both cases, the greatest number of participants were found in the theft scene (n = 14 in Siren; n = 9 in Togo) and disguise scene (n = 10 in both Siren and Togo). Two facets combined the efforts of a considerable number of participants in these scenes. Within the theft scene, the rental-fraud facet required more participants than the various parking lot thefts. In Siren's disguise scene, amateur body switchers were greater in number than professional body switchers. In the Togo operation, the inverse was the case.

The main brokers in each case were not necessarily found in scenes that contained the most participants. Two brokers in the Siren network were located in the market scene (Nodes 1 and 44), with three others found in the theft (Node 11), conceal (Node 6), and disguise (Node 15) scenes. The market scene was also a main segment for brokerage in the Togo network, with two top brokers (Nodes 1 and 16). Two other top brokers were located in the disguise scene (Nodes 2 and 29) and another in the conceal scene (Node 3).

Aside from Node 11 in the Siren network, and as expected by Tremblay, Talon, and Hurley (2001), brokerage was not a feature of scenes located at the extremes of the crime-commission process—namely, the theft and disposal scenes. It may seem obvious that the extreme scenes in a crime script are less likely to contain participants with considerable brokerage capital because of an absence of either a preceding or subsequent scene. But the incorporation of a network perspective within the script framework reveals that the exclusive script outlook may be misleading because participants in either the theft or disposal scenes in the Siren and Togo operations were linked beyond adjacent scenes (namely, the conceal and market scenes). Node 11's emergence as a key broker in the Siren network should therefore not be taken as an anomaly. Furthermore, brokerage in the multiple facets in the disposal scene may emerge if we had information extending beyond the people that were responsible for importing the vehicles in different countries. Knowledge of subsequent links (buyers and sellers of the stolen vehicles once they arrive in each country) will definitely increase their brokerage capital within the overall ringing scheme.

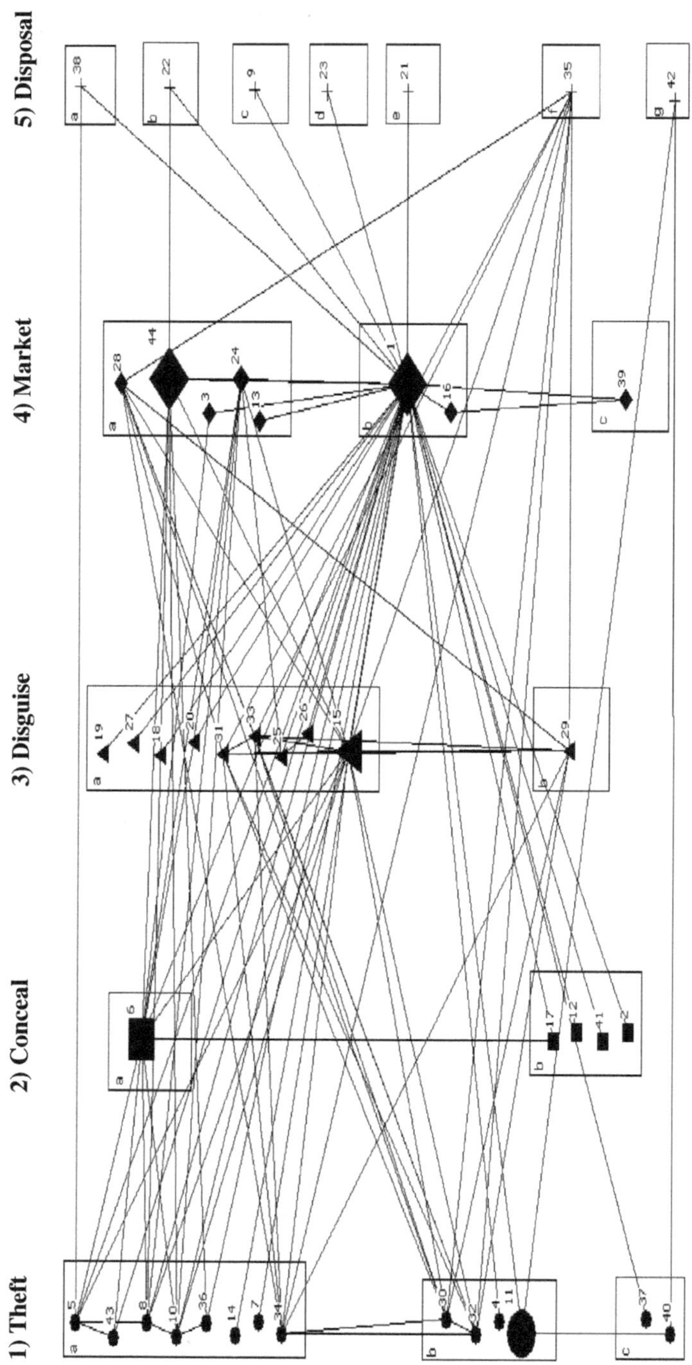

Fig. 7.1 The network within the Siren script

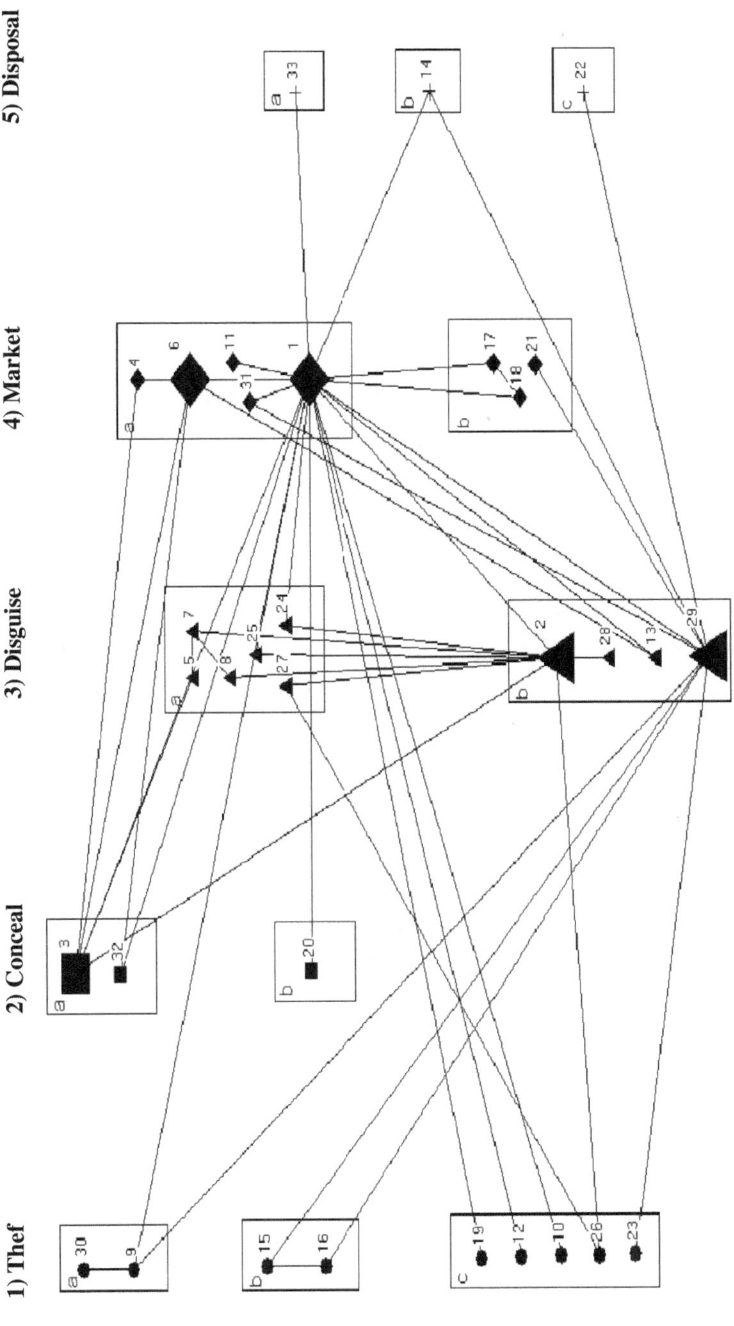

Fig. 7.2 The network within the Togo script

There was no linear process in either network-script representation (recall Ekblom and Tilley's "computer game" analogy). Participants in the theft scene were not exclusively connected to participants in the conceal scene; participants in the conceal scene are not simply connected to participants in the disguise scene; and so on. There was also some connectivity between participants within each scene. The only exception was found in the disposal scene, where importers in various countries were not in contact with each other. Thus, although not linear processes, paying attention to how the script was designed around certain participants did flesh out the links through which the diverse channels of action were executed within the ringing operations.

General Participant Removal and Impact

In either case, the removal of single participants would not result in the complete downfall of the crime script. In the Siren case, the removal of single participants revealed the following outcomes (recall that the network consisted of 44 participants and the script was comprised of 252 initial combinations):

Thirty-two participants (73% of the network) had no impact on the number of combinations;
Eight participants (18%) reduced the number of script combinations by 14%;
One participant (2%) reduced the number of script combinations by 33%;
Two participants (5%) reduced the number of combinations by 50%;
One participant reduced the number of combinations by 90%.

In the Togo case, the trend was similar (recall that the Togo network was made up of 33 participants and the script offered 72 initial combinations). Single participant removals resulted in the following outcomes:

Twenty-six participants (79% of the network) had no impact on the number of combinations;
Four participants (12%) reduced the number of combinations by 33%;
One participant (3%) reduced the number of combinations by 50%;
One participant reduced the number of combinations by 56%;
One participant reduced the number of combinations by 67%.

In each case, less than five (or less than 10% of) participants would reduce the script's flexibility by a significant margin (50% or more). Key players were therefore present in each operation. The subsequent analysis pursues this issue within the brokerage framework.

The Broker Impact on Flexibility

Whereas the transaction process representing the ringing networks may appear entangling, there is a form of order that does emerge. If any coordination is in place in either script, it is through the brokers who play a crucial part in keeping the facets within the script's order in place. Although extreme cases are a factor here, average brokerage measures for the ensemble of participants in each scene reveal that such coordination is more likely from the market scene followed by the disguise scene. In the Siren case, the mean for betweenness centrality scores for participants was under 1% for the theft, conceal, and disposal scenes, 3% for the disguise scene, and 10% for the market scene. Mean brokerage leverage was lowest for participants in the disposal and theft scenes (0 and 2, respectively), 12 for the conceal and disguise scenes, and 129 for the market scene. In the Togo case, mean betweenness centrality varied from under 1% in the theft and disposal scenes, 3% in the conceal scene, 6% in the disguise scene, and 10% in the market scene. Mean brokerage leverage was, once again, lowest in the disposal (0) and theft scenes (0.44), 5 in the conceal scene, 13 in the disguise scene, and 42 in the market scene.

With substantial variation in brokerage capital observed and with three to five participants from different scenes in the script emerging as clear brokers in the network, the next step was to evaluate what impact the removal of the top brokers would have on the script process. The argument is that brokers in the crime script are pivotal for increasing the number of options that the criminal network has for executing the various scenes and for achieving a common objective.

Table 7.2 illustrates the multitude of permutations that were available within the Siren and Togo ringing networks. The impact of brokers within the Siren and Togo scripts can be examined by assessing change in the maximum combinations for all scenes within the scripts before and after key brokers are removed from the networks.

The Siren operation was made up of a more elaborate script and a larger network than the Togo operation. As Table 7.2 demonstrates, broker extractions had an important impact on both operations.

Using betweenness centrality as an indicator, the top three brokers in the Siren case were S1, S15, and S11. As mentioned earlier, Freeman's betweenness centrality measure and Gould and Fernandez's brokerage leverage coefficient were strongly correlated. However, the difference between

Table 7.2 Script permutations and combinations after removal of top three brokers

	Manipulation criteria	Script permutation	Maximum combinations
Siren	*Betweenness Centrality*		
	All Participants	$3 \times 2 \times 2 \times 3 \times 7$	252
	Remove S1	$2 \times 2 \times 2 \times 1 \times 3$	24
	Remove S1, S15, S11	$2 \times 2 \times 2 \times 1 \times 3$	24
	Total Brokerage Leverage		
	All Participants	$3 \times 2 \times 2 \times 3 \times 7$	252
	Remove S1	$2 \times 2 \times 2 \times 1 \times 3$	24
	Remove S1, S15, S6	$2 \times 0 \times 2 \times 1 \times 3$	0
Togo	*Betweenness Centrality*		
	All Participants	$3 \times 2 \times 2 \times 2 \times 3$	72
	Remove T1	$3 \times 1 \times 2 \times 2 \times 2$	24
	Remove T1, T29, T2	$0 \times 1 \times 2 \times 1 \times 0$	0
	Total Brokerage Leverage		
	All Participants	$3 \times 2 \times 2 \times 2 \times 3$	72
	Remove T1	$3 \times 1 \times 2 \times 2 \times 2$	24
	Remove T1, T29, T2	$0 \times 1 \times 2 \times 1 \times 0$	0

the two was sufficient to identify a different participant amongst the top three brokers. S11 is replaced by S6 amongst the top three brokers if Gould and Fernandez's brokerage leverage is used as an indicator.

If we were to remove S1, the main coordinator within the market scene, the maximum number of script combinations drops from 252 to 24 possibilities (a decrease of 90%). The most damage would occur within the disposal scene, since four of the seven importers were exclusive contacts of S1. If we use betweenness centrality to designate the remaining two brokers in the top three, no additional impact is observed. However, the removal of S11 (rather than S6), based on the brokerage leverage measure, does have an additional impact in that contacts with participants in the conceal scene are no longer accessible. At this point, a scene is completely lost and the script is considered broken.

The impact of the top three brokers in the Togo case would have been much more detrimental. In this case, the top three brokers (T1, T29, T2) were designated as such by both betweenness centrality and brokerage leverage measures. The removal of T1, the top broker, would have decreased the script's combinations from 72 to 24 possibilities (a drop of 67%). The removal of the remaining two brokers in the top three would break the script at two important parts—access would no longer be available to participants in the theft and disposal scene. The impact is notable in that the remaining participants in the script would have to find access to others who could initiate and end the ringing operations.

Nonbroker Impact on Flexibility

It may be that brokers are more difficult to remove from a crime-commission process. They may be as resourceful in securing their position as they are in profiting from it. Some intervention strategies may therefore opt toward targeting a greater number of less-important participants who secure themselves to a lesser degree. In this section, this option is tested to see if such an impact would be as significant as that produced by the removal of the top three brokers.

Table 7.3 presents the results of three trial tests in which a random selection of 15 participants in the Siren and Togo cases were extracted (the top three brokers were excluded from this random selection). In the Siren case, the removal of this sizeable set of participants (about one-third of the network) would result in a script that, although damaged, remains functional. Trials 1 and 3 resulted in a decrease of 71% from the initial 252 combinations. Trial 2 was somewhat more damaging with a decrease in combinations by 86%.[4] The Togo network followed a similar pattern—some damage, but

Table 7.3 Script permutations and combinations after removal of random 15-node sets

	Trial: (random set)	Script permutation after random set removal	Maximum combinations
Siren	1: (2, 4, 5, 9, 10, 12, 13, 14, 20, 21, 23, 29, 31, 32, 41)	3 × 2 × 1 × 3 × 4	72
	2: (3, 7, 9, 12, 16, 17, 21, 22, 26, 28, 29, 35, 37, 40, 44)	2 × 2 × 1 × 3 × 3	36
	3: (2, 5, 9, 13, 14, 16, 17, 27, 29, 30, 35, 36, 38, 43, 44)	3 × 2 × 1 × 3 × 4	72
Togo	1: (3, 6, 8, 9, 13, 15, 18, 20, 21, 22, 23, 24, 27, 30, 33)	2 × 1 × 2 × 2 × 1	8
	2: (3, 4, 5, 8, 9, 11, 18, 19, 20, 21, 24, 25, 28, 31, 33)	2 × 1 × 2 × 2 × 2	16
	3: (5, 6, 8, 12, 14, 15, 16, 17, 18, 20, 22, 23, 25, 27, 30)	2 × 1 × 2 × 2 × 1	8

[4]Note that S29, who was selected in all three random trials, represented one facet on his own in Siren's disguise scene. He was the auto-industry professional who entered the network because of his expertise in windshield removal/replacement (see discussion above).

no breakdown. Trials 1 and 3 decreased the script's flexibility (from an initial 72 combinations) by 88% and Trial 2 decreased the maximum number of combinations by 78%. In either case, the ringing scripts would still be operational and the impact of removing the three top brokers in each network would be more important than that of removing 15 (and maybe more) of the remaining participants combined.

V. Conclusion

The ringing operations studied in this chapter were centralized and resilient because they contained brokerage features that increased the level of flexibility for achieving the collective goal. Removing the main brokers would have decreased this flexibility to the extent that each crime script would be no longer functional.

Which scenes in a script are sensitive to brokerage removal varies from network to network. The loss of a conceal scene, as was the case in the Siren script, would be less consequential to a criminal network than the loss of theft or disposal (or both) scenes, as was the case in the Togo operation. Finding someone else who could hide stolen vehicles for a temporary period would appear to be less difficult than replacing participants in the extreme scenes. In this sense, the Togo network was more sensitive to broker removal than the Siren network.

Merging script and network frameworks benefits each side of the partnership. An exclusive network approach would identify breaks in the structure of coparticipation, but would not be able to assess if the criminal operation was still functional. Thus, an important challenge for social network analysts seeking to disrupt dark, covert, or criminal networks through the removal of central participants of various kinds (see, for example, Borgatti 2006) would be to address whether a fragment from a newly broken network still contains all the necessary scenes of the crime-commission script. If the new fragment still contains all the elements of the script, breaking the network has not had much of an impact on the crime-commission process.

Overlooking the network approach, at the same time, is also limited in that an exclusive script approach would reveal the more vulnerable segments of the crime-commission process, but without an understanding of the people executing such action, we would have difficulty assessing a strategy for disrupting the process. Combining the two approaches is important since the script and network representations in our study did not follow the same logical order.

One last issue extending from the findings from this chapter's case study, as well as those in the previous chapters, concerns the high correlations between brokerage measures and degree centrality. In the Siren and Togo networks, as with the Ciel and Caviar networks, the same participants were high on both degree and betweenness centrality. In such cases, the brokerage capital provided by these participants did contribute to increasing the level of flexibility in each network, but these participants were also making themselves more vulnerable by exposing themselves with a high degree of direct connectivity. As the next chapters will illustrate, in some networks, the distinction is clear between hands-on participants and those occupying brokerage positions.

Chapter 8
Hells Angels in Springtime

If the idea of flexible order is truly pertinent in the context of crime, it must be applicable in network settings that integrate established criminal groups and formal organizations. In the Montreal and wider Quebec context, there is no better case for this than the network surrounding the Hells Angels. It would be difficult to tell what crime trends and patterns would have been like in Quebec over the past three decades without the Hells Angels. Although not a criminal organization per se, the motorcycle club's members and associates have been so intrinsic in the province's criminal markets that they have had not only a key role in shaping the structure of such markets, but also in determining the policies and controls that were developed to contain organized crime in such activities.

As often happens in such contexts, the image of the Hells Angels as a social threat can be so consuming that, during certain periods, the club has become the embodiment of organized crime, particularly in political, law-enforcement, and media circles. As a result, Naylor's (2002, 1997) wise advice not to *confound an association of criminals with a criminal association* (p. 40) is lost in the midst of repeated claims that the province's criminal markets (and primarily illegal drug-distribution markets) are controlled within an infrastructure governed by the six Hells Angels chapters across Quebec.

Naylor's warning is indeed sound. It is unlikely that the Hells Angels' formal organizational structure is transferable to a criminal market of any scope. However, extensive law-enforcement investigations and recent trials of a large number of the club's members and associates lend some credence to this unlikely transfer. The present case study uses material from these investigations and trials to examine this possibility. The line of inquiry guiding this chapter examines the extent to which a criminal network mirrors the formal organization in which many of its participants are members and associates.

C. Morselli, *Inside Criminal Networks*, DOI 10.1007/978-0-387-09526-4_8,

I. Hells Angels Inc.

At their official website, the Hells Angels qualify themselves as the *oldest and biggest 1% motorcycle club in the world.* The "one-percenters" status refers to those motorcycle clubs that are not registered with the American Motorcycle Association or the Canadian Motorcycle Association (see Barger 2000; Wolf 1991). With a history spanning more than half a century, the club has emerged into a vast organization, with chapters spanning across five continents. The organization's name and its skull logo are officially registered under the Hells Angels Motorcycle Corporation and protected by international law.

The onset and expansion of the Hells Angels in Quebec has been similar to their emergence elsewhere, but one feature that remains unique to the Quebec branch is the club's association with crime and violence from its arrival in the province during the late 1970s. Tremblay, Laisne, Cordeau, MacLean, and Shewshuck (1989) and Alain (2003) scanned the evolution of the one-percenters biker groups in Quebec and identified three phases since the early 1970s. In the first phase, from 1971 to 1978, most of this one-percenters subpopulation was located in rural Quebec, increasing from a population of roughly 600 to 900 members from various gangs. During this period, many alliances were formed between small biker groups. This period was also marked by the arrival, in 1977, of the first Hells Angels chapter (in Sorel, Quebec), which immediately established itself as an overriding organization of the smaller biker groups. During the second phase, from 1978 to 1983, 800 individuals were identified within the one-percenters classification. This period was marked by increased violence between clubs and by the positioning of members from several groups within various illicit drug markets (particularly for synthetic drugs). According to Alain (2003), the population of one-percenters in Quebec dropped to 300 during the final phase from 1984 to 2001. This period also highlights the presence of the Hells Angels as a dominant group within this biker subpopulation.

Although these studies provide a helpful count of the Hells Angels and other bikers with one-percenters status, a clear change in the club's reach must be established as of 1994, the year in which the sixth chapter was created in the province. This last chapter was a Nomad chapter that was comprised of the most-reputed Hells Angels members across the province. While Nomad chapters are not restricted to a specific geographical territory, this new group quickly became a heavy presence in the Montreal region. In Montreal, the predominant clubs preceding the creation of the Nomads included members of Hells Angels chapters from outside the city and a Montreal-based group, the Rock Machine. The Rock Machine was at the core of an amalgam of small biker groups and independent drug merchants who joined forces to form the Alliance during the latter half of the 1990s, largely in reaction to the overwhelming presence of the Hells Angels Nomads in the city's drug markets.

Between 1994 and 2001, the Hells Angels and Alliance were at the heart of a lengthy biker conflict that led to a substantial number of killings in and around the city. The claim in law-enforcement and popular media circles was that control of Montreal's illegal drug markets was at stake. Reports vary, but the most valid count maintains that, during this 7-year period, 261 victims were implicated in the confrontation between these two factions, leading to 126 murders and 135 attempted murders—55% of these victims were members or associates of the Hells Angels (for more details on these events, see Morselli, Tanguay, and Labalette 2008). Previous periods (most notably the late 1970s and mid-1980s) were also punctuated by important events and atypical levels in biker-related homicides, but no other period in Quebec or Canadian history has been marked by such consistent and clustered homicides over such an extended period of time and around a specific group.

The involvement of the Hells Angels in this lethal conflict increased their already public notoriety and their status as a prime target for law-enforcement controls in the province. Such attention was not new for the club. Fifteen-years earlier, members of the Hells Angels were amongst the first group to warrant the attention of a public commission that focused on the criminal activities of its members and the club's expansion across the province (Commission de police de Québec 1980). During the mid-1980s, the perception of the Hells Angels threat became important enough to warrant the contracting of a former hitman for the biker organization into the province's most controversial informant. The 7-year biker conflict between the Hells Angels and the Alliance was the most important in terms of its implications on organized crime policy and controls in both the Quebecois and Canadian context. In the fall of 1995, following the accidental death of a young boy who was struck by a fragment of a bomb that exploded and killed a Hells Angels' affiliate, the province's first specialized organized crime squad was put into operation to contain the escalating biker conflict in the Montreal region. This squad (or task force), dubbed the Wolverines, combined the efforts of investigators from the Sûreté du Québec, the Montreal Police, and the Royal Canadian Mounted Police. The main target of the Montreal Wolverines was the criminal activities of the Nomad chapter members, their underlings within the Hells Angels, and an affiliated group (the Rockers—not to be confused with the enemy gang, the Rock Machine) that was sponsored by the Nomad chapter and whose members were suspected of being the main executors of the Nomads' commands. The specific law-enforcement focus on the Montreal region ended during the summer of 1996, when the Wolverine squad expanded to include the Quebec City region. The

Montreal/Quebec Wolverine squad was active for another 2 years until May 1998, when it was decentralized and replaced with a province-wide infrastucture of six investigative squads (known as the Mixed Regional Teams), designed to mirror the six Hells Angels chapters across the province.

Aside from the changes that took place on law-enforcement approaches to organized crime, the Hells Angels threat also triggered the first anti-gang (or gangsterism) legislation in Canada. Canadian legislators had been traditionally reluctant to adopt such legislation in previous decades. Although public commissions from the late 1960s to the late 1980s had cast the spotlight on various criminal groups and organizations in illegal markets and legitimate industries, public policy was consistently restrained from mimicking the American neighbours who drafted their criminal enteprise (RICO) legislation in 1970 and applied it widely and with considerable impact throughout the 1980s and 1990s. This all changed with the Hells Angels threat of the last decade. The growing number of murders in Montreal's crime scene escalated to the point that citizens were increasingly exposed and sometimes victimized as bystanders to the biker conflict. In the spring of 1997, Canada adopted its gangsterism legislation. This legislation was modified less than 4-years later, in 2001, after two prison guards were murdered and a popular newspaper journalist was shot by individuals who were suspected to be following orders from Nomad chapter members.

The investigative efforts to contain the criminal activities of Hells Angels members and associates by the Wolverine squad and its expanded version, the Mixed Regional Teams, came to a sudden and successful halt at the end of March 2001, when close to 150 individuals across the province were arrested and charged in what was (and remains) *the biggest organized crime sweep in Canadian history* (Sher and Marsden 2004: 254). This major crackdown against the Hells Angels was named Operation Springtime (see Chapter 2). Criminal charges against those arrested ranged from weapons offences, money laundering, conspiracy, drug-trafficking, murder, and gangsterism. These arrests led to Canada's first maxi-trials that spanned into 2004, when the last set of accused pleaded guilty.

The evidence assembled against Hells Angels members and associates was massive. This included the colossal collection of electronic surveillance transcripts intercepted throughout the task force investigation that led to the Operation Springtime crackdown. Also included in the ensemble of evidence were the affidavits that described the overall workings of the organization and the implication of each arrestee therein. By this point, the description of the Hells Angels' inner workings had become common "knowledge," with the daily coverage of the investigation and trials constituting the prime focus for most media outlets. The general formulation beginning each affidavit prepared for the Operation Springtime arrests described the Quebec Hells

Angels, and particularly the Nomad–Rockers segment, as a tightly knit organization with a clear pecking order in its domination of the illegal drug-distribution activities of its members and their underlings. The prosecution's line of argument throughout the case was the following (this description originally appeared in French—the translation is mine):

- *The present investigation establishes that the members of the Nomads and Rockers organizations form one gang, in which all participate and contribute to an important extent.*
- *This gang exists only for the commission of lucrative crimes that serve to enrich the gang and its members. The most frequent of these crimes is illegal drug trafficking, but other crimes, such as intimidation, assault, use of explosives, conspiracy, and murder are also common.*
- *(. . .) The Hells Angels Nomads and Rockers carry emblems that they refer to as "patches" and that such emblems identify their membership status. (. . .) The wearing of patches or other objects that identify members in the motorcycle club are often used for purposes of intimidation and are proudly exposed by members and ambitious associates who aspire to gain official membership in the club. These symbols therefore serve as a mark of commerce and as tools of intimidation for facilitating the control of illegal drug selling territories.*
- *There exists, within the realm of the Hells Angels organization and its affiliated gangs, a well-established hierarchical structure and mode of function in which each individual has a role. There is also interdependence between members and the diverse crimes that they commit. Committing crimes under the Hells Angels banner has major advantages that are due to the criminal status and notoriety of the organization, the terror regime that it has developed and the rapport de force that it has demonstrated throughout recent years, as well as the national and international scope of the organization.*
- *All individuals who are part of this organization are sponsored by an official member and have to gain the approval of 100 percent of members in order to climb the hierarchy. All have to be useful to the ensemble of the group. Individuals at the lowest level of this structure serve those at the highest level—the contrary is never the case. Aspiring members climb the hierarchy in accordance with their utility, such as the volume of sales in illegal drugs, their contacts for importing illegal drugs, or their capacity to commit violent acts ranging from minor assaults to murder.*
- *For an affiliated gang, the structure of this organization is composed of Friends, Hang Arounds, Strikers, and Full-Patched members. For the Hells Angels, the structure is composed of Friends, Hang Arounds, Prospects, and Full-Patched members. In general, Hells Angels recruit*

> *members from affiliated gangs, but this is not an obligation in that some members do come from elsewhere. (...) The Full-Patched member of an affiliated gang must have the 100 percent approval of all other members of that gang. At this level, the member manages an illegal drug trafficking network, alone or with other members. He has acquired the trust of the group and commits, with or for them, different crimes. He takes part, at this level, in decisions that concern the group. The affiliated gang is also sponsored by a Hells Angels chapter and exists exclusively to serve and execute different tasks for members of this chapter, such as the distribution of illegal drugs and assuring their protection. All members and strikers must pay 10 percent of the profits generated from illegal activities to the Club. (...) Within a Hells Angels chapter, a Hang Around has a superior status than full-patched members of an affiliated gang. A Hang Around may order and direct them. A Hang Around must also serve members of the two superior echelons in the chapter. At this level, the Hang Around has already proven himself as a criminal and is well known within the group. He is sponsored by a Full-Patched member. The Hells Angels Prospect is also sponsored by a Full-Patched member and has to have the approval of 100 percent of members in order to climb the hierarchy and hope to become a Full-Patched member. At this level, he manages criminal operations in partnership with another member or association of members. He climbs the hierarchy in accordance with his effectiveness, availability, loyalty, and contribution to the group. The Full-Patched Hells Angels member is autonomous in the criminal activities that he manages alone or in association with others, but he must respect the rules and philosophy of the chapter. This level is the highest in the illegal drug distribution pyramid. His role is to supply drug distribution cells. At this level, he orders and commands crimes rather than executes them himself. It is a fact that no one could climb the levels of this hierarchy without committing a crime. The members of this gang are therefore all criminal.*

This well-regulated criminal system was the model professed by law-enforcement officials, prosecutors, and media outlets throughout the trials of the Hells Angels members and associates following the Operation Springtime crackdown in 2001. Key points that should be retained from this interpretation are that:

(1) The Nomad chapter of the Hells Angels and its affiliate clubs existed only for criminal purposes (specifically, illegal drug distribution) and recruited only offenders.

(2) Hierarchy was the governance model within the organization and of any criminal activities that extended from the organization.
(3) Climbing the echelons within the hierarchy was the motivational force driving all to participate in criminal activities.
(4) Top-ranked members were privileged in that they were able to order lower-level members, while remaining active in profitable criminal activities from a distant and secure position.

This description offers the claims against which this chapter's case study is designed. Using the electronic and physical surveillance records detailed in Chapter 2, the analytical sections examine the extent to which the Hells Angels hierarchy smoothly translates into the ensemble of interactions between its members and associates, as seen by the network analysis of communications between participants. Centrality measures help designate key participants in the overall network, but different types of centrality measures offer different interpretations of how a network is structured. On one hand, we may expect the highest-ranked members (the Nomads) to be amongst the most central, particularly in terms of the number of direct contacts they maintain in the network (degree centrality), their connections to heavily connected participants in the network (eigenvector centrality), or their ability to manage closely linked teams of participants (clustering coefficient). On the other hand, clustering measures and centrality measures based on direct connectivity (degree or eigenvector centrality) are more likely to indicate a participant's vulnerability to detection within the network.

Emphasizing the last point in the prosecution's line of argument—that the Full-Patched Hells Angels member *orders and commands crimes rather than executes them himself*—, we would expect the highest-ranked members to be relatively low in their direct connectivity in the network, thus confirming their ability to capitalize on others who manage and execute crimes in their place. This latter scenario, however, is not necessarily consistent with a straightforward command-and-order system. The prosecution's account also offered another interpretation in its emphasis on partnerships between members of different ranks in the organization. High-ranked members may have been active as partners in the criminal activities, but with different networking patterns than lower-ranked members. Centrality measures that capture a participant's capacity to broker between other participants represent strategic forms of networking. Such measures either account for a participant's capacity to mediate along the shortest paths uniting disconnected participants in the network (betweenness centrality) or to mediate less efficiently along alternative paths that unite disconnected participants in the network (flow betweenness centrality).

Thus, in the context of a criminal network structure, in which direct connectivity is an indication of visibility/vulnerability and brokerage is a more subtle and strategic positioning pattern, we should expect lower-ranked members of an organization to be more visible (more vulnerable) and higher-ranked members to be more brokerage-like (more strategic) in their networking patterns. Settling the issue of whether higher-ranked members are hands-on or perform strategic maneuvers in criminal networks (and inversely for lower-ranked members) will help us clarify not only the interpretation held during the prosecution of these Hells Angels members and associates, but also similar claims that appear in regard to the presence of organizational structures in criminal activities.

II. The Hells Angels Network in Operation Springtime

The analyses in this chapter examine the extent to which the organizational structure of the Nomad–Rockers organization is coherent with the communications that made up the network intercepted during the extensive task force investigation of these groups and the drug-distribution operations that surrounded them. Note that this case study is distinct from all other case studies in this book in that it is not based on individual criminal ventures. In short, this is not an action-set network. Instead, the drug-distribution operations studied here are best described as an ensemble of action-sets that were linked to the Nomads and Rockers and targeted during the investigations that led up to the March 2001 crackdown. The Operation Springtime network is a representation of the ongoing criminal activities of a specific group of offenders across a wide range of ventures and transaction channels.

Figure 8.1 illustrates the intercepted communication patterns within and between the formal ranks that constitute the Hells Angels organization targeted during the investigations. The 48 participants who had formal membership status in the Hells Angels organization (black nodes) are distinguished from those who did not (white nodes).

Once again, formal rankings in the Hells Angels are as follows: *Full-Patched members* are at the top of the hierarchy; *Prospects* rank below full-patched members; and *Hang Arounds* or *Friends* rank lowest within the organization. The Rockers' organizational structure is similar to that of the Hells Angels: *Full-Patched members* had the highest status, followed by *Strikers*, *Hang Arounds*, and *Friends*. Regardless of their rank within the organization, Rockers are subordinates to all members of the Nomad chapter.

To follow the formal ranks, the graph in Fig. 8.1 is best examined from the core outward. Some may argue that a graphic representation that illustrates the informal networking of members and associates within the formal hierarchy making up the organization should follow a vertical display. Figure 8.1 does

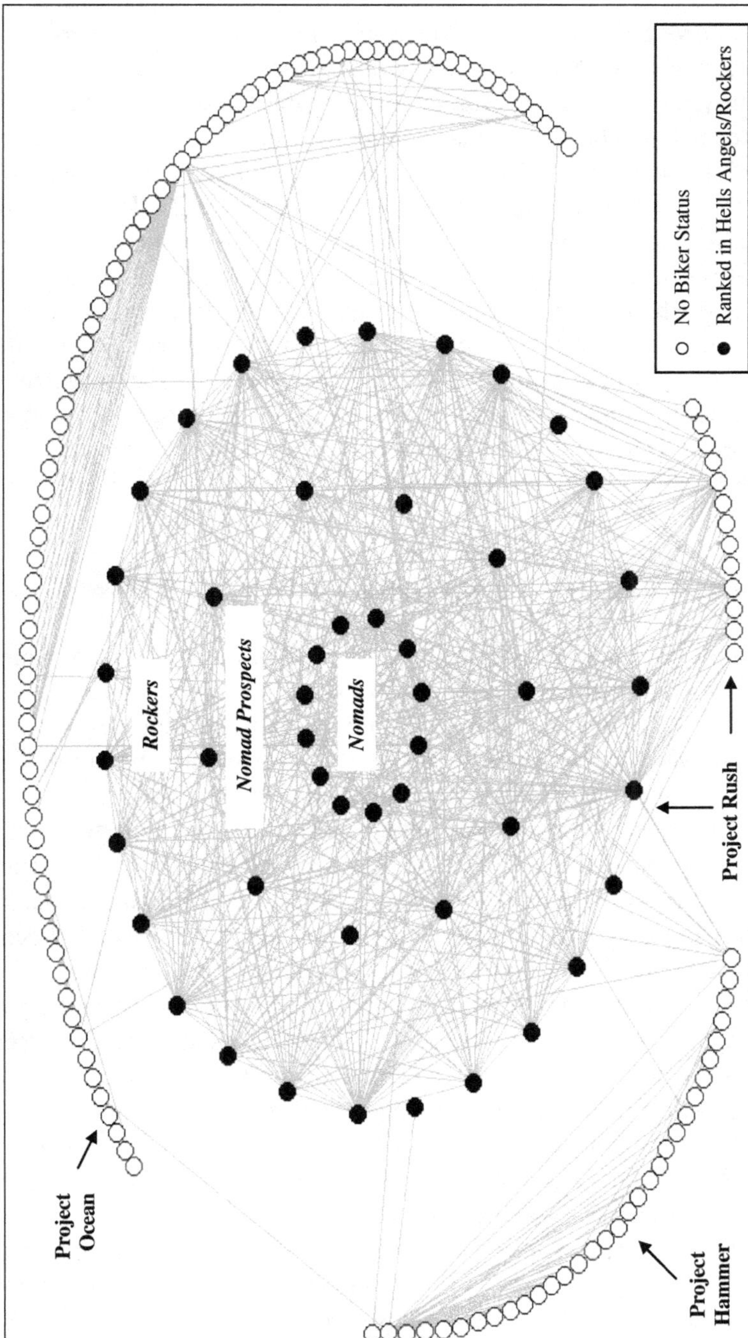

Fig. 8.1 The Hells Angels network in Operation Springtime

present such a vertical display—just imagine that you are looking at a pyramid from above. The highest ranked Hells Angels—the 12 members of the elite Nomad chapter—are presented at the core. The layer surrounding the Nomad members shows the ten Nomad Prospects. The third layer presents the 26 participants who had an official status with the Rockers. The outer layer of the network presents all participants targeted during the investigations but who did not have formal membership status in either the Rockers or the Hells Angels.

Figure 8.1 also illustrates how participants were targeted across the investigations that led to Operation Springtime (see Chapter 2). The 48 participants with formal ranks in the Nomads/Rockers and 13 participants without formal status (grouped at the bottom-center of the graph) were targeted during Project Rush. The 81 nonbikers that span the upper to right side of the graph were targeted during Project Ocean, which focused specifically on the financial routes extending in (profits) and out (costs) of the organization. The 32 participants positioned at the bottom-left of the graph were targeted during Project Hammer, which focused on a specific cocaine-trafficking group that extended from the Nomad core.

Variations in Key Participant Status Across the Network

If a bias taints this data, it would be because most of the intercepted communications centered on the Nomads and Rockers who were at the center of the investigations that generated this case study's main data sources. Nomads, in particular, were the most-heavily targeted during the investigations. Thus, if there was a fundamental law-enforcement bias guiding the present analysis, we would expect Nomads to be amongst the most central participants, in terms of direct connectivity. This is not the case and the relative importance of Nomads as central participants varies. Figures 8.2, 8.3, and 8.4 illustrate how degree, eigenvector, and betweenness centrality are distributed across the overall network (larger nodes indicate more central participants).

Figure 8.2 presents the network with an emphasis on those participants who had higher degree centrality. Aside from a few participants scattered across the Project Ocean and Project Hammer portions of the network, most participants with a high degree of direct contacts held an official rank within the Hells Angels organization. However, few of the Nomads had high degree centrality. This pattern was more concentrated amongst the Nomad Prospects and even more so amongst the lower-level Rockers.

It may be, then, that Nomads were letting their prospects and the Rockers do the hands-on work for them—at least within the scope of intercepted communications. If Nomads were simply positioning themselves close to lower-level members who maintained a high degree of direct contacts in the

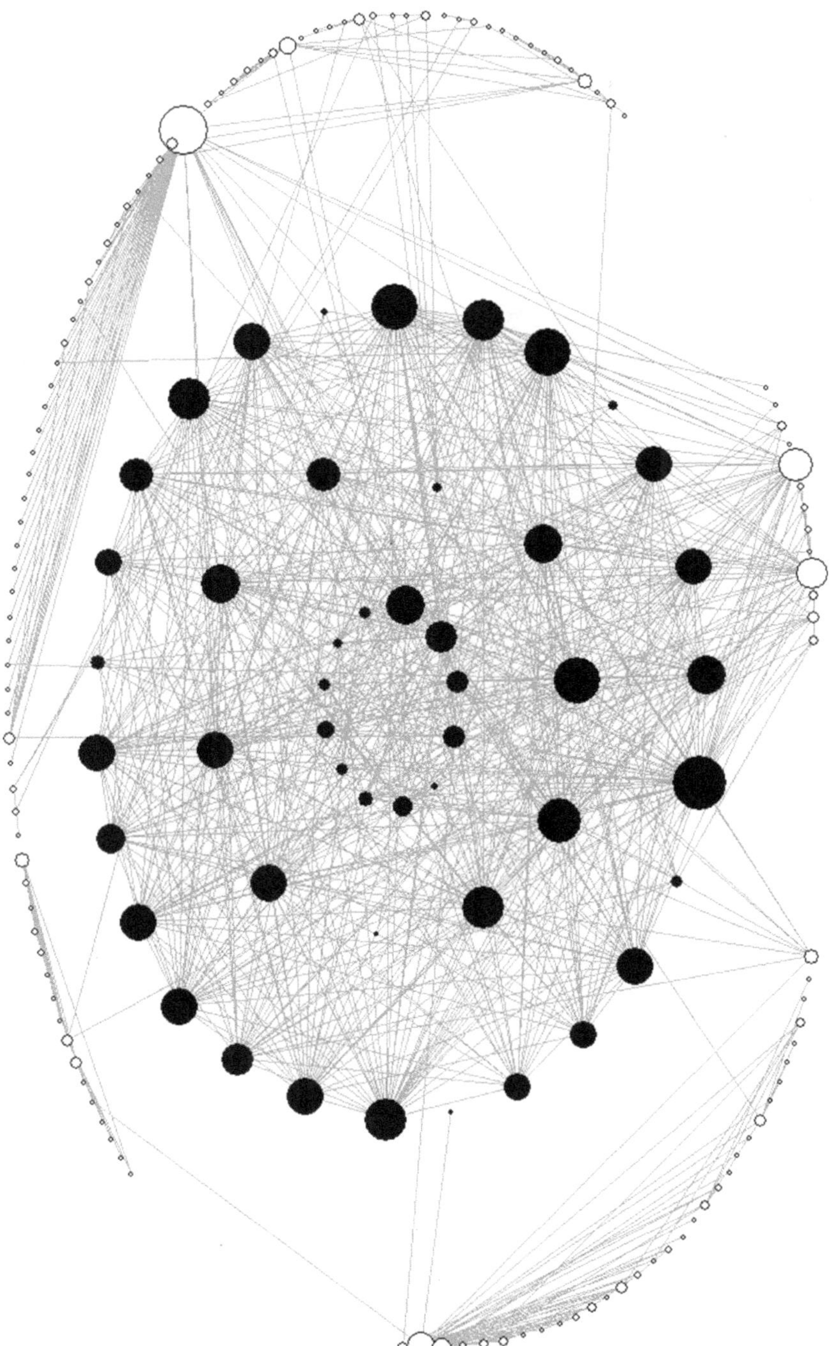

Fig. 8.2 The Hells Angels network in Operation Springtime (key participants designated by degree centrality)

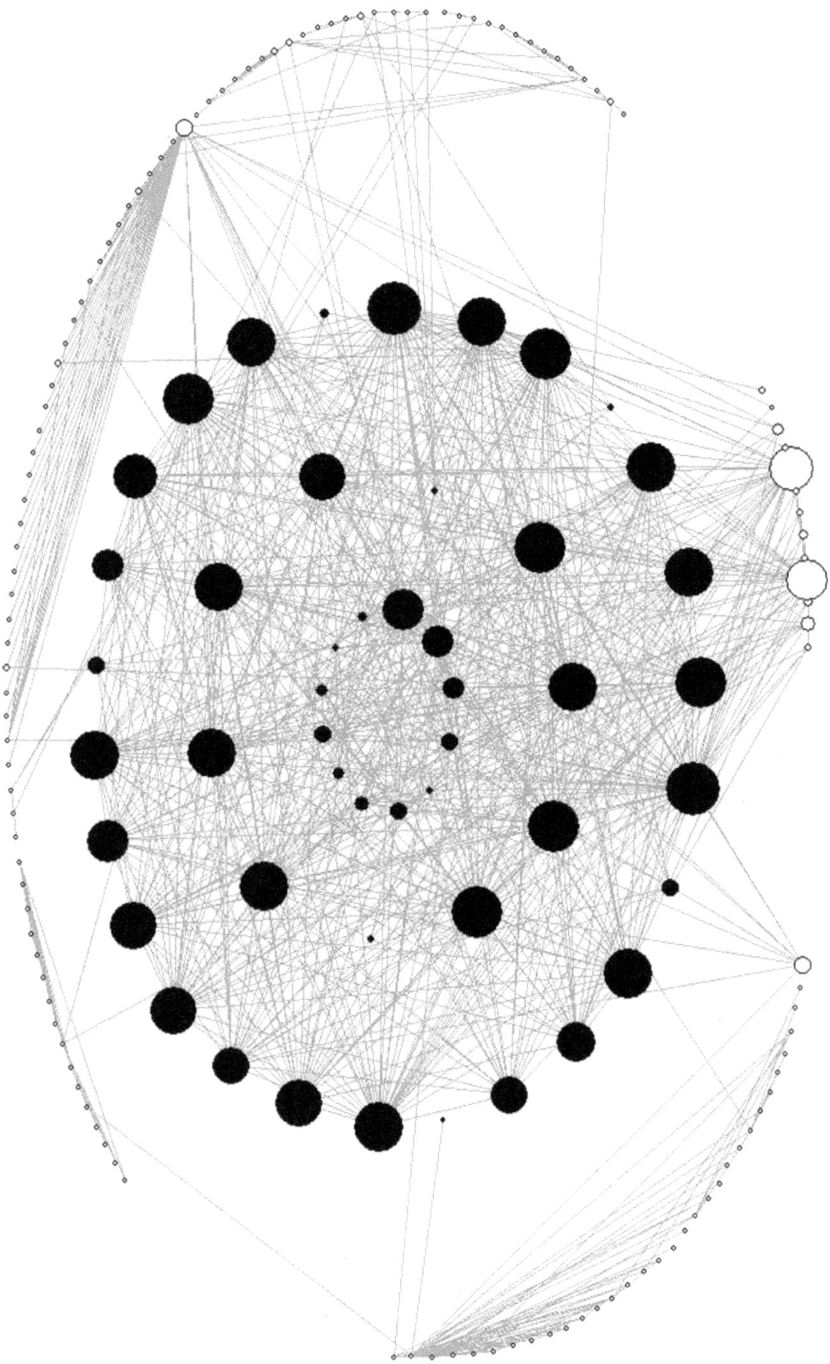

Fig. 8.3 The Hells Angels network in Operation Springtime (key participants designated by eigenvector centrality)

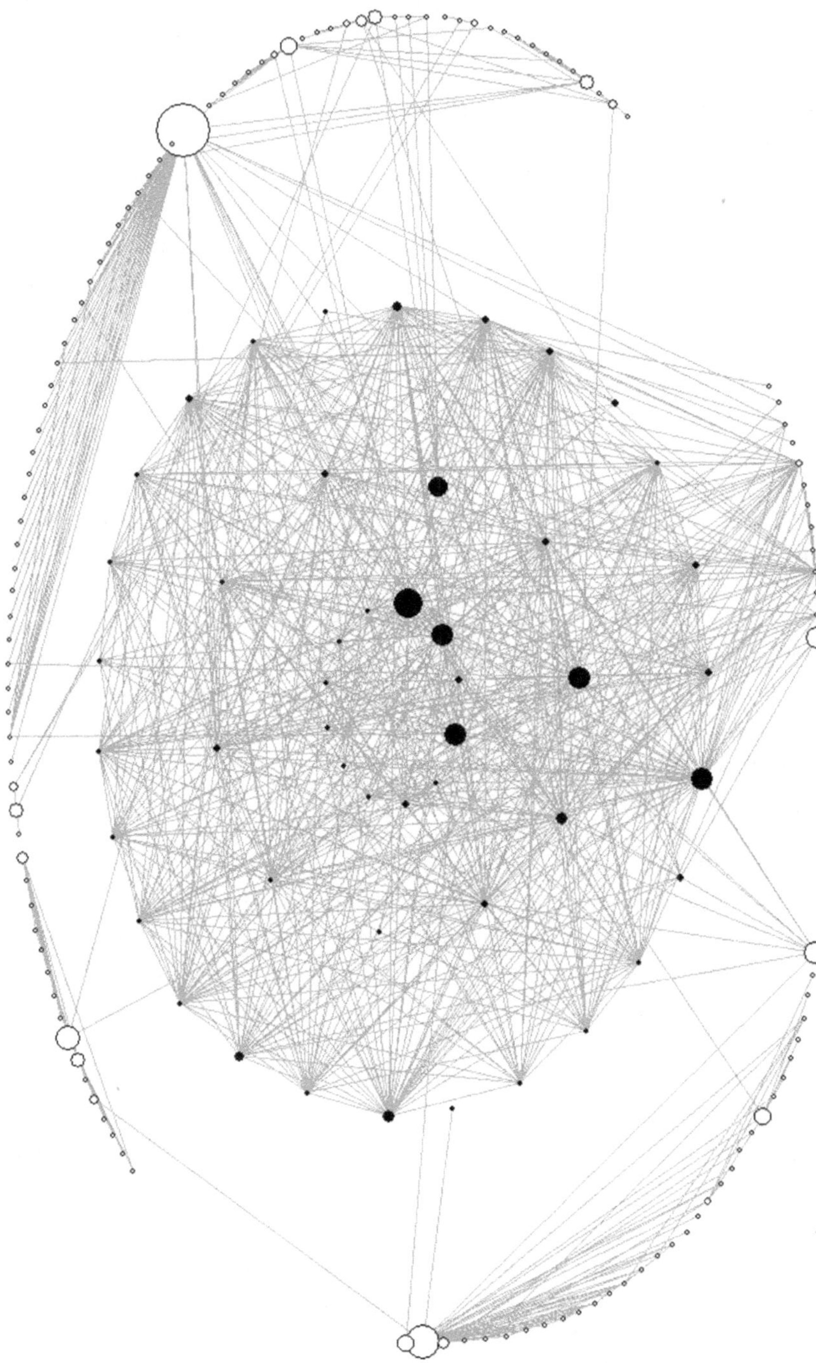

Fig. 8.4 The Hells Angels network in Operation Springtime (key participants designated by betweenness centrality)

network, we would expect such patterns to emerge in an analysis of eigenvector centrality. But, once again, this is not the case. Figure 8.3 presents the network with an emphasis on those participants who had higher eigenvector centrality. The patterns are similar to those found in the previous illustration depicting the degree centrality distribution, although the eigenvector centrality distribution is more heavily concentrated amongst those participants with an organizational status within the Hells Angels. Nomad members are, once again, less prominent in eigenvector centrality than Prospects and Rockers.

If there is a common link uniting all the case studies in this book, it is that brokerage capital is a fundamental networking pattern in criminal networks. This was the case for the freelancing and partnership drug-importation networks studied in the first analytical chapters as much as it was true in the networks built for stolen-vehicle exportation. Figure 8.4 shows that brokerage is a pattern that is also more particular to Nomads in the Operation Springtime network—at least more so than for lower-level members in the Hells Angels. The participant with the highest betweenness centrality was a nonbiker who was targeted during Project Ocean. Other participants with relatively high betweenness centrality were targeted elsewhere in Project Ocean and during Project Hammer. However, within the formal ranks of the Hells Angels (those targeted during Project Rush), the key participant status that was so vividly associated with the majority of Rockers and Nomad Prospects when assessing the degree and eigenvector centrality distributions substantially diminishes. Nomads, on the other hand, are amongst the few participants within the Hells Angels organization that are also identified as key participants from a betweenness centrality outlook.

Table 8.1 summarizes the patterns emerging from the previous figures.[1] Clustering coefficient scores and flow betweenness centrality scores are also added to the previous centrality measures. These findings are telling in that average scores for the distinct ranks in the Hells Angels organization reflect the patterns that were typical of members in their networking. The two extreme ranks in the organization networked in inversed patterns—members of the Nomads and Rockers are always at opposite ends of the continuum for each of the measures. Whereas Nomads were relatively low in direct connectivity (degree centrality, eigenvector centrality, clustering coefficient) and relatively high on brokerage-like connectivity (betweenness centrality, flow betweenness centrality), Rockers were high in direct connectivity and low

[1] The results in Table 8.1 must be considered in the structural make-up of the overall network, which, according to the network centralization results (bottom row of table), were relatively low for all centrality measures. Thus, the analysis involves a search for central participants in a network that was not heavily centralized to begin with.

Table 8.1 Means for network characteristics across Hells Angels organizational ranks

	Direct connectivity			Brokerage connectivity	
	Degree	Eigenvector	Clustering coefficient	Betweenness	Flow betweenness
Nomads (n = 12)	7.95	7.56	57%	3.56	0.58
Nomad Prospects (n = 10)	16.13	20.89	73%	3.18	0.49
Rockers (n = 26)	15.30	20.11	76%	1.30	0.37
Non-gang members (n = 126)	2.15	0.78	63%	1.35	0.94
Network centralization	22%	25%	66%	31%	10%

Note: All actor centrality measures are normalized.

in brokerage connectivity. Consistent with the formal organizational structure, Nomad Prospects find themselves somewhere in the middle of the elite Nomads and the bottom-level Rockers—they were high on all forms of networking.

Although the diverging patterns are consistent with the formal ranks within the Hells Angels organizational structure, these results do not necessarily represent a hierarchically defined command-and-order system. As a group, members of the Nomad chapter were older (44-years of age, on average) than Nomad Prospects and Rockers (36-years of age for either subgroup). Criminal career experience and criminal maturity is therefore relevant and the relationships between a biker's age and the network measures are consistent with those found between formal ranks within the Hells Angels organization. Within the ensemble of participants who held membership status amongst the participants targeted during the investigations making up this case study, age was negatively correlated with degree centrality, eigenvector centrality, and the clustering coefficient. Age was positively correlated with betweenness centrality and flow betweenness centrality.

III. Conclusion

The formal ranks in the Hells Angels organization do emerge in the communication patterns studied in this chapter—the network, in short, did mirror the hierarchy. However, the network was not necessarily "behaving" as a hierarchy. If we take formal rank within the organization to be an indication of prestige or reputation, most of the higher-level gang members in this network were relatively low in the number of direct contacts they maintained, low in the direct contacts they maintained with others who had a high degree of direct contacts, and low in the density of their personal clusters of contacts in the whole network. The more reputed members of this criminal network were therefore not "hands-on" participants. They were, instead, indirectly involved, as suggested by their higher brokerage capital (betweenness centrality and flow betweenness centrality).

The results in this study do not clarify the causal order between networking and formal ranks. Two scenarios are possible. The first scenario reflects the law-enforcement and prosecutorial description of the Hells Angels' implication in the targeted drug-distribution activities. In this description, Nomads had a great deal of authority and control over higher-ranked members and ranks within the organization defined how a member could position himself amongst others—in this sense, formal ranks define the network structure and members receive network privileges after reaching higher promotional levels. The second scenario would follow a tournament setting, as suggested by Levitt and Venkatesh's (2000) analysis of a Chicago gang's financial structure. If this is the case, promotions are an extension of what a member has been proven to be capable of—strategic networking patterns should therefore precede promotion and a good indication of who the up-and-comers are within an organization's ranks and the network that mirrors it will be those lower-ranked members who network through patterns similar to high-level members. Rank, in this sense, is but a prize for knowing how to fit in amongst others and not a formal authorization to govern the actions of others.

Chapter 9
Street Gang Presence in Drug-Distribution Operations

It did not take long for Montreal Police officials to publicly announce that another growing crime threat was facing the city after a considerable portion of Quebec's Hells Angels were arrested, accused, and incarcerated during the Operation Springtime case. The new threat came in the form of street gangs that had emerged over the previous decade from Montreal's ethnic communities. In this chapter and final case study, the flexible order thesis and brokerage focus is carried into a portion of Montreal's street gang landscape.

Street gangs have been of concern to the Montreal police and public since the mid-1990s, when they were believed to be made up of youths who were operating prostitution rings and working as satellite drug-distribution groups for the more prominent adult criminal organizations in the city, such as the Hells Angels. By 2001, the main youth gangs of the 1990s, such as the Montreal-North based *les Bo-Gars* (the *Handsome Boys* in English) and the *Crack Down Posse*, had grown up into young adult gangs. In 2007, the Montreal Police maintained that there were 20 major street gangs operating in the city. For several years, these groups and the general street gang population have been systematically classified under one of the two clashing supra groups from past Los Angeles gangland: the Bloods (*les Rouges* in French) and the Crips (*les Bleus* in French). The Bo-Gars, for example, are considered to be part of the Bloods, while the Crack Down Posse are considered to be part of the Crips. Emphasis on this dichotomy in police and media circles has created an overall sense of polarization and a climate of fragile tension amongst observers of Montreal's street gang scene. No study, however, has provided strong evidence to support this image of the street gang landscape in the city.

The presence of such groups in criminal activities has also been the subject of much attention by the police and media. Claims have been repeatedly made that street gangs are growing more organized, more vicious, and more dominant in criminal activities, such as drug-trafficking. However, and as found elsewhere, the street gang landscape in Montreal is not likely to fit

C. Morselli, *Inside Criminal Networks*, DOI 10.1007/978-0-387-09526-4_9,

into what Felson (2006) refers to as the "Big Gang Theory." This "theory" refers to the tendency of some researchers, policy makers, law-enforcement officials, and popular outlets to centralize the gang phenomenon around one or a few groups, when, in fact, such settings are more likely populated by many small groups, some of which could be perceived as full-fledged gangs and others amounting to nothing more than a small group of troublesome youth.

Another problem with the Big Gang Theory is how it applies to criminal activities. Approaching crime from a gang standpoint distorts our understanding in that the gang is not the most suitable analytical point for making such an appraisal. Gangs and gang members, like other offenders, participate not in gang networks, but in criminal networks that include an assortment of members from different gangs and a variety of participants without any gang membership status. Of course, the presence of gang members in some criminal networks could be marked by a level of dominance that supports contentions that such structures are, by and large, gang networks. However, the extent to which a criminal network is governed by a gang presence is an analytical issue that can be addressed most adequately within the network framework that has been under development throughout this book.

The case study that is presented in this chapter begins with an examination of past research on gang structures and gang presence in criminal activities. As we will see, the network framework has been repeatedly upheld as a relevant approach for understanding gang patterns and for generally downplaying the stereotypes and threat claims that often accompany this phenomenon. The main findings from past research are carried over into the case study analysis that centers on three investigations of drug-distribution activities conducted by the Montreal Police. These drug-distribution operations were believed to revolve around one of the older and more reputed gangs in the Montreal landscape—*les Bo-Gars*. The case study is meant to assess whether the presence of this gang, or other gangs whose members were implicated in the network, was indeed an important structuring force within the drug-distribution operations. This case study pursues a similar analysis to that in the previous chapter, with the main line of inquiry addressing how members from fixed groups or organizations fit in the communication patterns making up a criminal network.

I. Gangs and Flexible Order

Coughlin and Venkatesh (2003) mark the mid-1980s as the period during which the "gang and drug problem" became synonymous in popular discourse and media and law-enforcement members began to systematically associate street gangs with drug distribution and other criminal market trades (p. 43). In their research review, the authors find support for the increasing implication of street-gang members in underground economies, but the scope and organization of such involvement varied considerably across major American cities, leading to much disagreement amongst researchers in regard to the form and importance of the gang phenomenon.

Research on street gangs resembles research on organized crime in many ways. The two fields are often approached as separate areas of study, largely because many organized crime researchers believe that gang research is primarily focused on youths and many gang researchers believe that organized crime groups are more sophisticatedly organized than the typical street gang. However, the questions and debates pursued by researchers from either side are very similar. For example, researchers in both fields are faced with the task of providing the necessary nuances to past endeavors that became the scientific basis for many of the stereotypes that are often associated to either phenomenon. Yablonsky's (1962) *The Violent Gang* is to gang research what Cressey's (1969) *Theft of the Nation* is to organized crime research. Although many facts, concepts, and accurate propositions could be extracted from either study, both contained too many exaggerations, misleading views, and a flair for dramatic narration that would come to be the basis for popular and law-enforcement views on such problems and would oblige researchers to dispel the many stereotypes over decades. Such stereotypes generally exaggerate patterns of violence and claim too much control of criminal activities in the name of a given gang/criminal organization. Thus, like organized crime researchers, gang researchers have focused considerably on the types and scope of organization within gangs, between gangs, and within larger criminal activities.

Such stereotypes are generally at the root of how serious a gang phenomenon is perceived. There is much research that explores the "threat" issue and the divide separating the competing outlooks has been reviewed by enough students of gangs that a complete review in this chapter would not add anything new. Once again, the division is similar to that found in the field of organized crime. One argument dismisses claims of a gang threat and emphasizes the moral panic climate that is often generated not by gang activity but by a misinformed emphasis on such a threat by law-enforcement officials, political actors, and media members (Archbold and Meyer 1999; McCorkle and Miethe 1998; Jackson and Rudman 1993; Huff 1990; Zatz 1987). The opposing argument is also critical of law-enforcement and political approaches to the gang phenomenon, but, contrary to the moral panic stance, officials are criticized for not doing enough to contain and control a growing gang problem—they are essentially in denial (Weisel 2002a; Huff 1990).

Whether law-enforcement, policy actors, and media members are exaggerating or avoiding a gang problem is a matter that cannot be explicitly addressed in this chapter. However, the overall climate which hovered over the law-enforcement investigations that underlie this chapter's case study was a clear example of the moral panic/denial dichotomy. To a considerable extent, the Montreal Police exaggerated the street gang threat in the city, particularly since the declaration that street gangs were filling the criminal market voids left by the incapacitated Hells Angels came within months of the Operation Springtime crackdown. At the same time, the Montreal Police was faced with the plight of having to do something about what appeared to be an evident (albeit unfounded) observation in regard to the city's criminal markets—street gangs were taking over where biker gangs left off.

What I argue in this chapter is that, rather than exaggerating or denying the state of the problem, such issues can benefit from an analysis that assesses how gang members participate in criminal activities and establishes the extent to which such participation constitutes a pattern that is unique to gang presence and that therefore requires specific attention and intervention for such a setting. Past research would suggest that this is not the case and that gang members, when interacting amongst themselves, or in the context of wider criminal activities, follow the patterns that are conducive to participation in the constrained settings in which criminal networks emerge. Gang members, in other words, are not different in their networking patterns when participating in criminal ventures.

The general assessment from most research reveals the loose structures in which gangs and gang members interact (Klein and Maxson 2006; McGloin 2005; Decker, Bynum, and Weisel 1998). Although some rare gangs could be relatively large and corporate-like (see Venkatesh and Levitt 2000; Levitt and Venkatesh 2000 for what appears to be a rare example), the actions of gang members are best approached from a clique or even an individual-member outlook. Even in contexts with a rare presence of a single dominant gang, the difference between the existence of such a gang and the actions of its members as a collective must be taken into consideration. Research as early as Thrasher's (1927), Short and Strodtbeck's (1965), and Suttles' (1968) studies of Chicago gangs confirmed that, aside from confrontations with other gangs, individual gang members typically act beyond the realms of the gang. Thus, a gang could have many members, but this does not necessitate that comembers interact with each other in a given setting. Cliques and resource sharing may emerge around the activities of some members, but aside from offering a common identity and franchise, the constitution of the overall gang itself is generally amorphous. Later studies lend additional support for such early observations (see Spergel 1995; Sanders 1994; Vigil 1988).

Klein is one of the researchers who contributed most to our understanding of the constitution of gangs. Over a period of 40 years, he and various collaborators studied the extent to which gangs and gang settings are cohesive. In an early study, Klein and Crawford (1967) argued that measures of cohesion based on an approach that accounted for members' interactions were more accurate tools than traditional measures based on verbal responses made by gang members and associates to a researcher (p. 70). Their approach incorporated a team of researchers who observed interactions between core and fringe members of a single gang over a 6-month period. These interactions were coded in a sociometric matrix that contained the daily observations recorded by social workers (or, as the authors referred to them, "detached workers") who were part of an intervention program for gang members in Los Angeles. These daily records permitted the creation of this internal gang network within the social work setting because observers recorded how many times they saw any given gang member and to what frequency that gang member was in interaction with other gang members. What Klein and Crawford essentially constructed was a valued, symmetrical matrix of gang member interaction. What they found in analyzing this matrix was that not all members in a single gang were in equal interaction with each other. Instead, clear subgroups emerged from the network representation that was derived from the social workers' daily observations and less than 50% of gang members were found to be part of the identified subgroups. Also, these subgroups were found to be more cohesive in their internal interactions between members than the overall gang and largely based on age differences between members.

That gangs were less cohesive than expected by the idea of a "gang subculture" has been repeatedly addressed in research following Klein and Crawford. Fleisher (1995), for example, maintained that the Crips *versus* Bloods division was often blurred during his fieldwork in Seattle, with members from "opposing" sides often found in amicable contact with each other. Another feature that is often associated with gang organization in popular and law-enforcement circles is that of leadership. But contrary to popular beliefs, leadership in a gang setting is more likely transient and versatile. Klein and Maxson's (2006) assessment of this is consistent with the flexible order outlook: *Most gang scholars have found leadership to be functional, shifting, unstable, and shared among many gang members. It often depends less on physical strength or criminal prowess than on verbal skills, opportunism, social capacities, and—in the case of traditional gangs—various age levels. Except in speciality gangs, leadership is usually not the hierarchical, command-oriented positional concept stressed by popularizers of gang matters* (p. 195). In the words of one of Klein's (1971) early interview respondents: *We got no leaders, man. Everybody's a leader, and nobody can talk for nobody else* (p. 96).

This is the individualized context in which gang matters have been consistently found to be acted out. Gangs, as Weisel (2002b) found in her own fieldwork with law-enforcement officials and gang members in Chicago and San Diego, are adaptive, organic, and flexibly structured so as to respond to the volatile environments in which they operate. Weisel demonstrated this by studying what were believed to be the four most-organized gangs in these two cities. What she found was consistent with an organic-adaptive organizational model: *[G]angs studied here feature the attributes associated predominantly with this form of organization. These include an emphasis on individual goals concurrent with organizational goals, diffuse leadership, the active role of subgroups, a generalist orientation, persistence in a volatile environment, and continuity despite the absence of hierarchy. Adaptive or organic organizations thrive in a volatile or changing environment, and organizations that survive under such conditions are more likely to maintain multipurpose and flexible structures, with flexible leadership and little differentiation among member roles* (p. 52). This account is consistent with Skolnick, Bluthenthal, and Correl's (1993) description of Southern Californian gangs of the 1980s as horizontally structured. These authors added that factors reflecting an individual's maturity (age, experience, and knowledge) were the most decisive features accounting for a gang member's status in the drug trade and within the gang setting (p. 213).

Low cohesion, volatile leadership, elder–novice relationships, and overall flexible order in gang settings have been concurrent aspects in past research on this phenomenon across a wide array of geographical contexts. There is not much about the Montreal context that suggests that matters should be different than that found elsewhere. There are, however, some questions that have been minimally inquired upon and that extend directly from these consistent findings. Whereas gang research has traditionally focused on interactions and organizational features within a strict gang setting, few have reframed the research designs to assess how gang members interact with each other and with nongang members in a wider criminal context. We know that gangs facilitate the criminal prospects of their members and associates (Thornberry, Krohn, Lizotte, Smith, and Tobin 2003; Fagan 1990, 1989). However, as studies from both a moral panic and denial stance have argued, we have to explore deeper into the extent and form of gang presence in criminal activities to be able to assess to what point a gang is a facilitating, coordinating, or dominating feature in the criminal activities of its members and nonmembers. The lack of cohesion and volatile leadership that are generally associated with gang settings suggest that domination by any one or few members over the actions of other members should not be expected. Both facilitation and coordination, however, are compatible with the shifting and flexible organizational structures generally found in gang settings. The extent to which gangs serve as mere facilitators or active coordinators in the scope of criminal activities may be addressed by focusing more specifically on how gang members participate in a given criminal network and to what extent their presence is a structuring feature.

II. Past Research on "Gang Networks"

The social network approach has been used by gang researchers in various ways for quite some time. Papachristos (2006) was right in maintaining that gang research translates naturally into a network framework. Past research that has focused on the network features of gang settings suggests that we should expect many small gangs interacting in a loose network rather than one or two large gangs overpowering a general milieu. This loose network representation is consistent with non-network research in revealing the lack of cohesiveness, transient leadership, and the overall flexibility of gang settings. Unfortunately, these common links have not been carried over when interpreting some of the more interesting research results extending from network applications in street gang settings.

The more ambitious projects that have applied the network framework have focused on gang-related homicides. The first research design to include this approach was part of the Boston Gun Project, a problem-solving research initiative that was designed to confront increasing trends of violence amongst youths in that city. In addressing this issue, Kennedy, Braga, and Piehl (1997) report their results from field interviews and mapping exercises conducted with street gang experts amongst Boston police officers, probation officers, and street-workers. The fieldwork was designed to arrive at a representation of the social networks of alliances and conflicts between street gangs in the Boston landscape. Centrality measures were used to identify key gangs in the ensemble of homicides and to orient intervention strategies to control the problem. Degree centrality and eigenvector centrality were used to assess the ensemble of gang conflicts. The authors reported that those gangs that scored highest with the eigenvector centrality measure were a closer match to those referred to as the more troublesome gangs during interviews with the various practitioners. No explanation was offered for why such a measure, that illustrates that the more troublesome gangs were those that were in contact with the most central gangs, was found to be a better indicator—the authors simply concluded that these gangs would make better targets for intervention.

Other researchers began to build on where the Boston Gun Project left off. Tita, Riley, Ridgeway, Grammich, Abrahamse, and Greenwood (2003) applied social network analysis to assess various strategies to reduce gang-related violence in Los Angeles' Hollenbeck district. Turning to a mix of law-enforcement officials and community group members, the authors derived a network representation of rivalries and violence amongst 29 street gangs in this area. Results illustrated how individual and collective attacks were largely contained within a geographical space divided by the San Bernardino Freeway. Rivalries crossed the freeway only by one bridge within the network that was created through the positioning of one gang, the *Hazard* (p. 11). Aside from the visual portrayal, no network features were analyzed in the report and nothing was made of this single bridge in the gang rivalry network.

Another research that was inspired by the Boston Gun Project took place in Newark. McGloin (2005) reported the results of her study conducted in collaboration with the New Jersey Gang Task Force. Her research began with a search for hot spots in gang activity and interactions in Newark and its surrounding area. A series of 32 focus group sessions with police experts who identified gang members and their respective associates revealed that four large gangs (the *Bloods*, the *Crips*, the *Latin Kings*, and the *Netas*) dominated Newark, but that they were dispersed across the landscape rather than concentrated in hot spots. Overall, there was no single encompassing gang network. As often found in previous gang research, cohesive subgroups (or cliques) were important elements that emerged from the analysis. McGloin pursued this finding by shifting the analysis to the individual level to assess whether specific members maintained any structural importance between these clusters of subgroups. Consistent with the lack of cohesiveness found in the gang-level analysis, only a handful of cut-points (gang members positioned as bridges) were found and identified as the only individuals connecting members between different gangs. In a later article that was based on the same data source and analytical strategy, McGloin (2007) revealed additional findings from this network analysis. Whereas degree centrality shed minimal insight into the behavioral patterns of gang members, betweenness centrality was significantly correlated with the arrest records and the number of weapons and drug-related offenses.

As with the previous two studies that applied network analyses to assess a gang setting, McGloin did not provide an interpretation of what her findings entailed for the Newark landscape. But there is common ground across the findings from these different studies in three different cities. In both the Boston Gun Project and the Hollenbeck study, the gang rivalry networks contained elements that breached the basic measure of degree centrality. Gang activities were not shaped around direct relationships. This was also consistent in McGloin's study, although her research was not framed within a specific criminal activity, as was the case with the gang-related homicide context for the previous studies. In the Boston setting, the more reputed gangs were those that were found to be in contact with other gangs that were more directly central (eigenvector centrality). In the Hollenbeck case, one gang kept the rivalry network in place within one component by bridging over a geographical divide. In McGloin's research, gang members who had greater experiences with criminal justice controls were those that had network positions reflected by higher betweenness centrality. These results are all consistent with the general appraisal from past gang research that gang structures are not cohesive. They are also consistent in that in loosely structured networks, degree centrality is rarely an important indicator for individual positioning. Alternative centrality measures that reveal the bridging and brokering capacities of key participants were designed to fit the needs of loose networks.

Thus, general research on gangs and applications of social network analysis in gang settings brings us to the following expectations. First, a gang is not a cohesive social entity and overlapping with other gang and nongang members should be expected. Second, subgroups within the gang are just as (or more) important as the overall gang itself. Third, gang settings are often organized around brokerage or cut-point patterns. But how do gangs and gang members that lack cohesiveness and position themselves in bridging patterns fit within the scope of various criminal activities? Klein and Maxson (2006) provide us with a cue for understanding how members of various types of gangs fit within the context of criminal activities. They found that larger gangs had lower arrest rates (number of arrests per number of members in the gang) than smaller gangs. Their explanation for this finding was that larger gangs tended to be more versatile in their crimes than smaller gangs, leading larger gang members to *engage in many activities which are of relatively little concern to the police* (p. 183). Although I cannot confirm to what extent Klein and Maxson's reasoning is accurate, taking into consideration the results from network research on gang settings suggests an alternative explanation for such a pattern. The alternative hypothesis would suggest that members of the more reputed gangs were those who were less directly central in gang activities, thus leading to lower arrest probabilities than members of smaller gangs. Rather than proposing that more reputed gang members are simply involved in less serious criminal activities, this alternative explanation proposes that the more reputed gang members are less likely to be hands-on participants when it comes to more serious criminal activities—instead, they participate from a distance. The versatility of the larger, more reputed gang members likely reflects their brokerage patterns.

That gang members are not directly central in gang settings is not a contradiction in itself. Katz, Webb, and Schaefer (2000) found that nongang members are often central and important participants in settings that are initially believed to be structured around gangs. Of course, to pursue this finding and test the extent to which this hypothesis is applicable requires a different approach than that generally found in gang research. A common aspect of past studies that designed a gang network through interviews with law-enforcement officials or data available in law-enforcement settings is that researchers generally began with a sample of gang members and proceeded to create a network between these members based on a variety of co-offending records, family ties, and surveillance observations. What these studies illustrate is how gang members are organized within a strict gang setting (e.g., how members are connected to each other; how members are connected within gang-related homicides). Instead, and as proposed here, we have to be able to observe gang members in interaction, but within a wider criminal setting, one that allows us to study the structure of criminal activities and assess gang presence therein.

III. Three Investigations, One Network

The case study in this chapter combines data sources from three investigations conducted by the Montreal Police (see Chapter 2). While it was not clear from the start of this research, these three investigations were progressively revealing a single criminal network of 70 participants that was organized around the distribution of drugs in Montreal North. These drug-distribution operations were believed to be under the control of one of the more reputed gangs in Montreal, the Bo-Gars.

There was a clear connection between the three investigations that were at the root of this case study's data sources. Observations made during Investigation A led to the onset of Investigations B and C. These latter investigations were partially connected by a common drug-distribution chain. However, at no time across these law-enforcement investigations was there an attempt to understand which participants were pivotal in linking the various drug-distribution activities across the three investigations. The focus throughout the investigations was on the gang members, and particularly the Bo-Gars members. The investigations were intended to disrupt the Bo-Gars, their suspected affiliate groups, and individuals who gravitated around this main gang. But while the principal concern triggering the first of these investigations was that Bo-Gars members and their affiliates were in control of the drug-distribution market in this sector of Montreal, no attempt was made to establish who were the key (or less central) participants in the overall network that was targeted from one investigation to the next and, more importantly, to what extent such key participants were members of the Bo-Gars or one of their suspected satellite gangs. The investigations, in short, were designed to target gang members as they participated in the distribution of drugs rather than target drug distribution and assess the place of gang members therein. Because of this, the investigations, although connected by network traces, were largely treated as independent cases when establishing who the central participants were and how gang members fitted into the organizational structure. As the analyses in later sections will illustrate, there are distinct features that emerge when examining a representation of the investigations as three single cases (and therefore three independent networks) *versus* a representation of the three investigations as overlapping to reveal a single network in continuous flow.

Identifying Gang Members

Once the participants in the drug-distribution operations were established across the three cases, gang members were identified in accordance with the Montreal Police's gang list. The Montreal context is consistent with experiences elsewhere. Katz and Webb (2006) illustrated how, throughout the past two decades, specialized gang units proliferated across North America and their main contribution was the collection, processing, and dissemination of intelligence data regarding the street gang landscape in various cities. These authors found that gang units that invested more heavily in intelligence gathering were also less likely to arrest known and suspected gang members, doing so only when they were certain that an individual was involved in a crime. This was described as an effective change from less intelligence-based enforcement strategies that involved random arrests, citations for minor acts, and harassment of gang members and affiliates. A key tool within this intelligence-gathering process was the creation of gang member/associate lists.

Past assessments of such lists have maintained that the underlying law-enforcement information and intelligence of gang activity are problematic. As with Katz's (2003) police setting in Junction City, the Montreal police list included individuals classified as hard-core gang members, gang members, or gang affiliates. The latter could be a problem when assembling such data since gang affiliations are largely identified and judged as such during police investigations—an individual is listed as a gang affiliate simply because s/he was observed as such during police monitoring. Thus, for the analyses in this chapter, associates are excluded and the designation of gang status is restricted to members only.

The presence of gang members from one investigation to the next reflects the ongoing objective of the overall operation that targeted the drug-distribution activities in the Montreal North district. The principal aim was to remove the street-gang members and particularly the Bo-Gars members.

The percentage of gang members amongst the participants targeted increased from Investigation A (18% were gang members), to Investigation B (23%), and to Investigation C (39%). This increase was even more revealing for the presence of Bo-Gars members across the three investigations (4% to 7% to 35% from Investigations A, B, to C).

Single-Case Versus Overlapped-Cases Representations

The organization of the drug-distribution operations and the place of key participants therein vary in accordance with single case representations of the three law-enforcement investigations and an overlapping representation of these three cases. For the first illustration (Fig. 9.1), the network is divided into three components that illustrate how participants were targeted in the separate investigations (e.g., participants targeted during Investigation A were only allowed contact with other participants in Investigation A; participants targeted during Investigation B were only allowed links with others in that investigation; participants in Investigation C were only connected to other participants in Investigation C). For the representation of the overlapped investigations (Fig. 9.2), the network was reconstructed to breach the investigative point of view and include all relationships between all 70 participants that could be recorded through the electronic surveillance, physical surveillance, and co-offending records.

Figure 9.1 presents the three components of the overall network that were targeted during each investigation. Investigation A is located at the bottom left; Investigation B is on the far right; and Investigation C is in the middle. The temporal flow of the investigations is also reflected in this illustration (from left to right) in that Investigation A targeted drug-distribution participants who were operating roughly 2 years before those in Investigations B and C, who were in operation during the same period.

Participants with the highest betweenness centrality scores for each network component are emphasized by larger nodes in Fig. 9.1. In Investigation A, N5 was the main broker, as were to a lesser extent, N4, N14, N24, and N28. In Investigation B, N47, N55, N61, and N66 were the main brokers. In Investigation C, aside from N79, the other participants with relatively high brokerage capital were N76 and N78.

Ten of these twelve participants in the drug-distribution operations were arrested at the end of the separate investigations (N61 and N79 were not arrested). The loss of these key participants, however, did not cease the network activities. We know this in large part because the drug-distribution operations continued in this area of the city long after the first arrests were made in Investigation A. Remaining participants and new arrivals in the

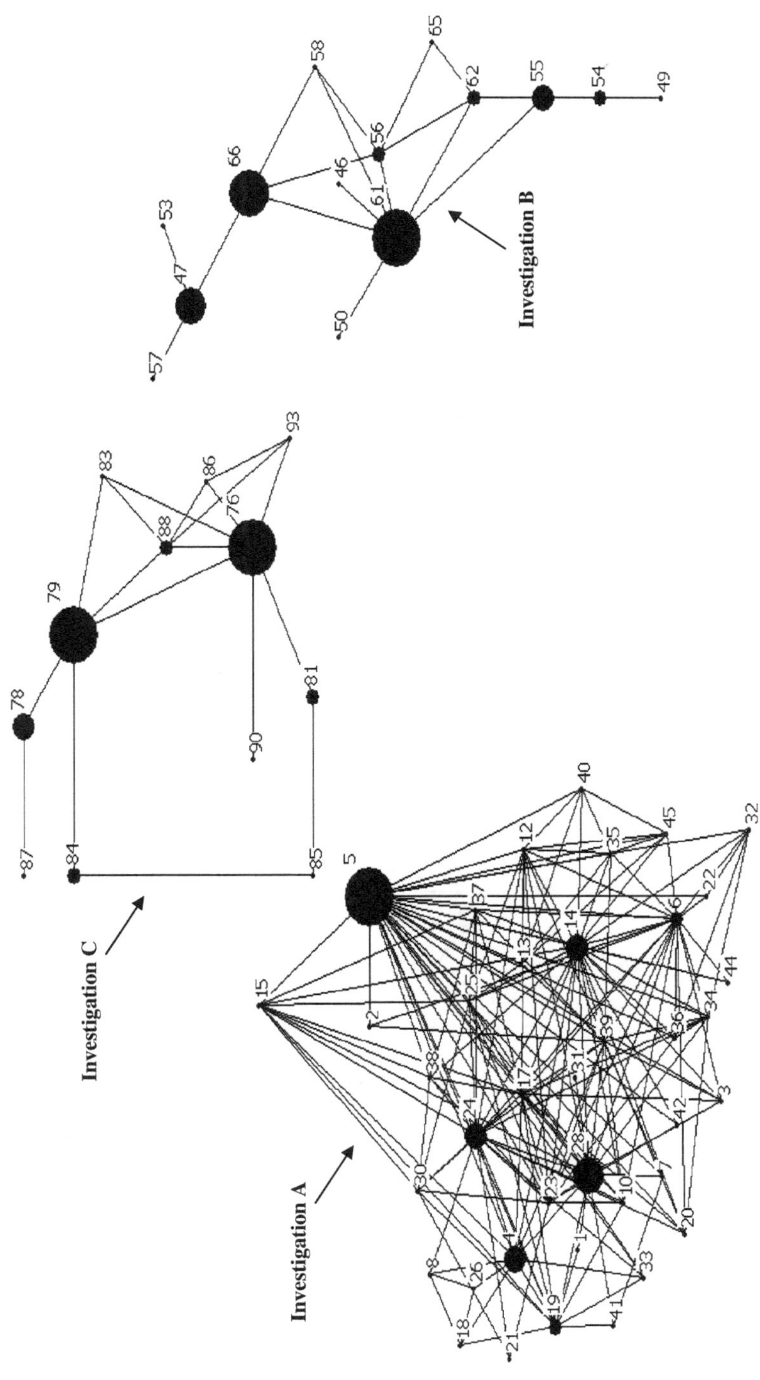

Fig. 9.1 Single-case representations of networks monitored during investigations A, B, C (key participants designated by betweenness centrality)

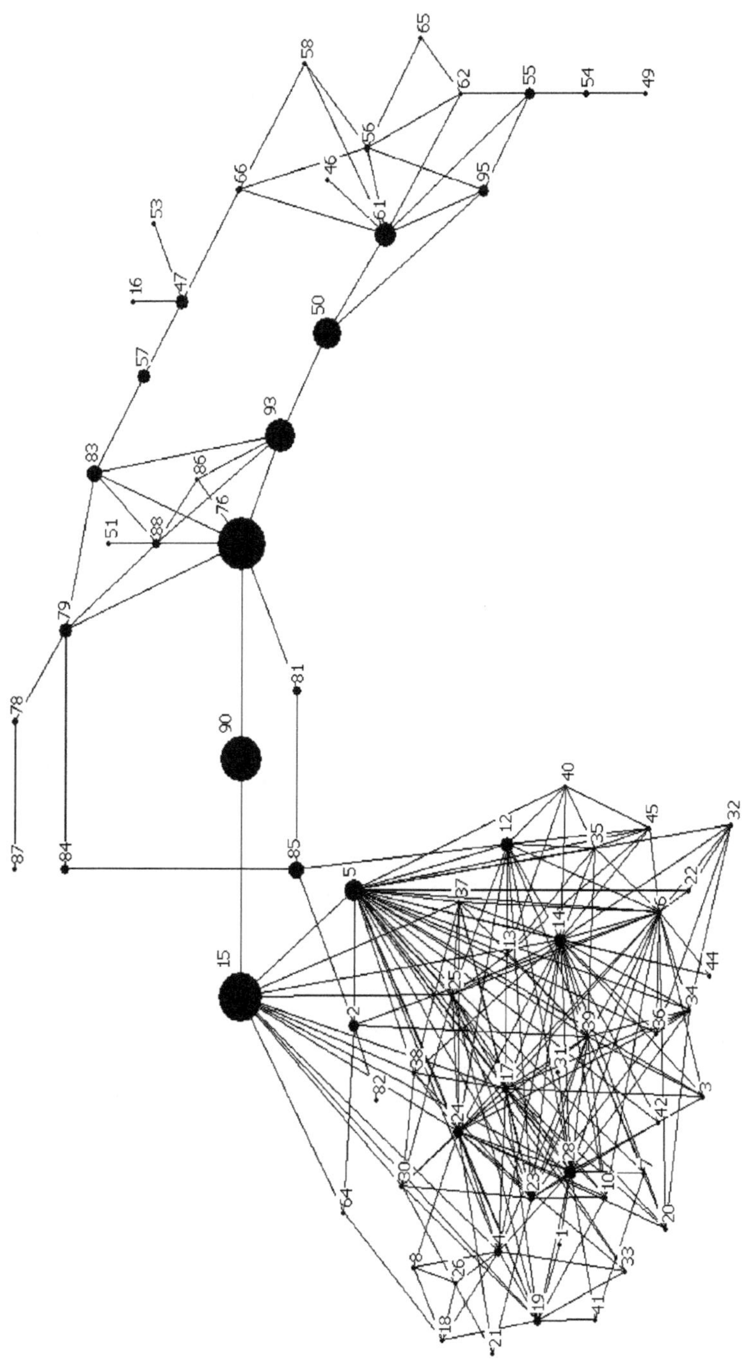

Fig. 9.2 Overlapped-cases representation of networks monitored during investigations A, B, C (key participants designated by betweenness centrality)

network adjusted to the loss of one or more key participants. Unfortunately, the representation offered in Fig. 9.1 does not allow us to see the remaining and new participants who kept the network in place after this first disruption. Such an analysis would require a representation that overlaps the components that are based on participants who were targeted separately across the three investigations.

This overlapped representation (Fig. 9.2) presents a single component and shows how the three investigations were indeed tapping into the same drug-distribution network. Two findings emerge from this new network representation. First, overlapping the investigations revealed five participants who were monitored in one investigation, but were also active in another. These five participants did not appear in Fig. 9.1 since they were isolated from the single-case representations. N16, who was first targeted (but not arrested) in Investigation A, re-emerged in Investigation C. N51 was targeted in Investigation B, but was more active in Investigation C. N64 and N82, who were targeted only, and respectively, in Investigations B and C, were already present as contacts to participants in Investigation A. N95, who was a key target in Investigation C, was more active in Investigation B.

Second, and most importantly, the transition from the drug-distribution operations that were targeted during Investigation A to those in the subsequent investigations is best observed by reassigning betweenness centrality scores within this overlapped network. The five participants with the highest betweenness centrality scores (N15, N90, N76, N93, and N50) in this network representation combine to create the *backbone* of the drug-distribution operations over the 3-year period. Aside from N76, none of these backbone participants were central in the earlier single-case representations of the drug-distribution operations (Fig. 9.1). Other participants should also be considered beyond their lower status as intermediaries in that, aside from the key broker from Investigation A (N15), N2 and N12 were also able to bridge toward N85 after the arrest of the key participants in Investigation A. These participants, and particularly N15, were the ones who came forward to help sustain the drug-distribution network regardless of the loss of key participants following the first of the investigations.

IV. Where are the Gang Members?

Since these three investigations had a mandate to target gang presence within these drug-distribution operations, many would (and did) expect that the most central participants would be gang members. This is not the case. Figure 9.3 adds the gang member emphasis to the previous illustration of the overall network that was targeted across the three investigations. Gang members are indicated by black nodes, while nongang members are indicated by white nodes. Also, any tie that involved members of the same gang are represented by thickened lines. Of course, not all members of the same gang were necessarily recorded as being in contact with each other in any of the Montreal Police's data sets used to construct this network. Members of a common gang are highlighted in Fig. 9.3 by the subset designations in the network.

Six gangs were represented in this network. The largest subset (across the upper portion of the network) was made up of 11 Bo-Gars members. Members from this gang were scattered across the network, with minimal ties recorded between comembers. The five members of the Young Master Crew are concentrated at the far right of Fig. 9.3. Although not clique-like, this smaller gang was more cohesive than the ensemble of Bo-Gars members. A clique was formed by three members of the Blood Game, at the lower center of the network. Although no contact was recorded between them, two members of the Bad Boys are positioned in the center of the graph. Finally, two single members from different gangs were also present in the network: N1 was a member of the Syndicate, an organization that was assembled in the late 1990s after a merge of several former juvenile gang members who became adults and continued to be active in various criminal markets. Police investigators who were involved in Investigation A, in which N1 was netted primarily through his contact with N28, suggested that his presence in the network was as a bridge to Hells Angels members and other organizations throughout the city who were active in wholesale distribution across a wider geographical territory (I had no way of verifying the validity of this claim). The second single gang member was N45, a member of the Dope Squad, who was also part of the clique comprising the three Blood Game members.

Indeed, the simple presence of 23 gang members, and 11 Bo-Gars members amongst them, would have many conclude that this drug-distribution network was governed by gang members. Some would also maintain that this network was dominated by Bo-Gars members. Three observations are offered to address these claims. First, only the Bo-Gars and the Bad Boys had members who were presented in more than one investigation, with the Bo-Gars represented in all three. Second, only one gang member (N50, a Bo-Gars member) was found in the set of five backbone participants who scored highest in betweenness centrality in the overall network (Fig. 9.2). Third, the remaining four backbone participants (N15, N90, N76, and N93) were known drug-traffickers. The basis of flexible order within this network was, thus, not within any of the gangs.

Regardless of whether nongang participants were associates who participated as representatives or subordinates of any one of the gangs found in this network, a more extensive analysis of this network does reveal a pattern surrounding the gang members, and the Bo-Gars members in

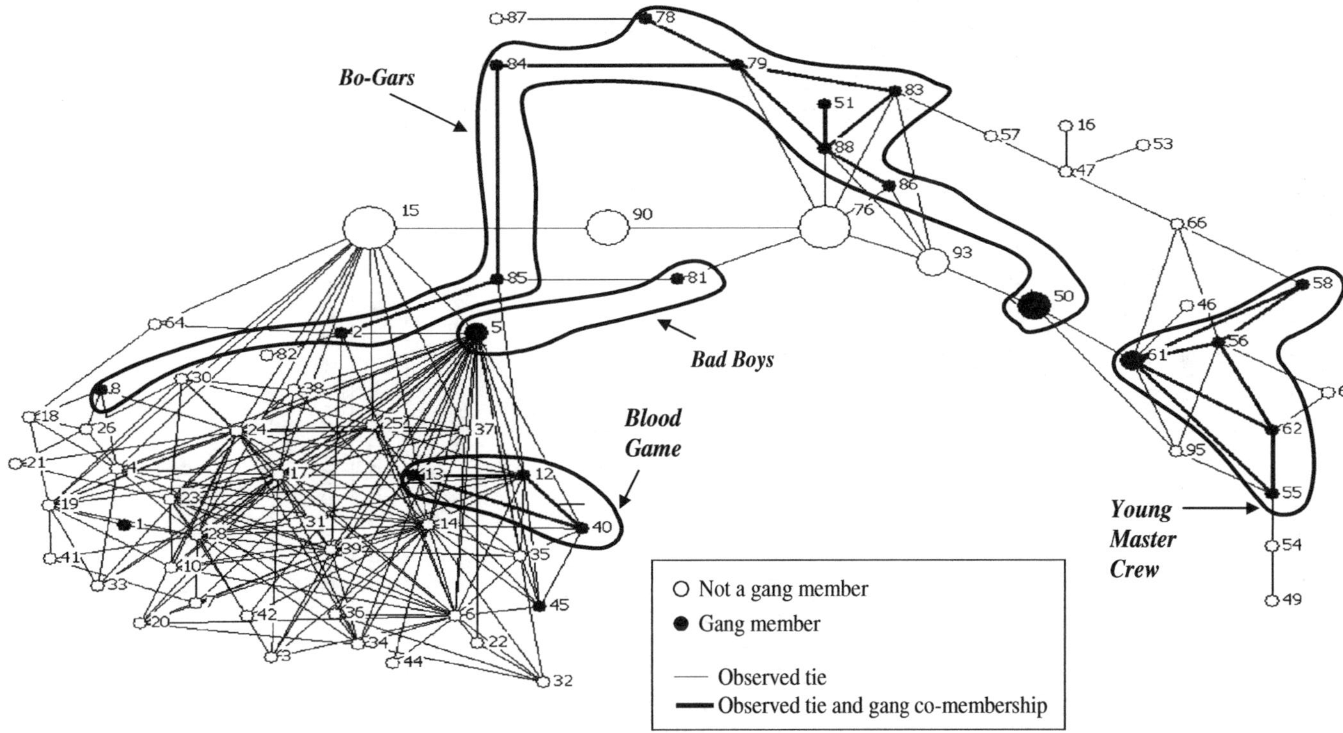

Fig. 9.3 Gang membership in overlapped-cases representation of networks monitored during investigations A, B, C

particular. Table 9.1 presents the results from a similar analysis as that conducted on the Hells Angels network in Chapter 8. Means for the various centrality scores are compared between nongang participants, Bo-Gars members, and other gang members in the network. These measures reveal how diverse gang members are (and are not) positioned in the overall drug-distribution network.

Table 9.1 confirms what was illustrated in Fig. 9.3—gang members were not more heavily connected than nongang members in the network. Degree centrality was similarly high for nongang and the smaller gang members. The Bo-Gars members, however, had less direct contacts in the network. Eigenvector centrality results show that nongang members had, on average, the most direct ties with the most heavily connected participants in the network. Although to a lesser extent than nongang participants, eigenvector centrality was also high for the ensemble of members from the smaller gangs. Once again, the major difference is found with the Bo-Gars, whose members proved to be much lower in eigenvector centrality. The third measure of direct connectivity within the network, the clustering coefficient, also confirms the distinctions between the Bo-Gars members and the remaining gang and nongang participants in the network. Indeed, the most likely participants to cluster were nongang participants, followed closely by members of the smaller gangs. Bo-Gars members were significantly lower in their clustering patterns. As Fig. 9.3 reveals, the Bo-Gars members were indeed present across the graph, but they were not closely intertwined with any particular set of other contacts. They were present across the network, but from a secure distance. Such presence-from-a-distance is particularly revealing for Bo-Gars members and ultimately placed into context with the betweenness

Table 9.1 Means for network characteristics across Bo-Gars members, other gang members, and nongang participants

	Direct connectivity			Brokerage connectivity	
	Degree	Eigenvector	Clustering coefficient	Betweenness	Flow betweenness
Bo-Gars members (n = 11)	5.4	1.6	35%	6.4	6.4
Other gang members (n = 12)	11.2	10.0	58%	4.5	2.1
Non-gang members (n = 47)	10.8	13.1	61%	4.4	2.4
Network centralization	35%	42%	56%	38%	17%

Note: All actor centrality measures are normalized.

centrality and flow betweenness centrality measures. The brokerage that is indicated by such measures best illustrate the network patterns of the Bo-Gars members.

The patterns found in this case study match those from the previous chapter. Like the Nomads in the previous case study, the Bo-Gars were more strategic and less hands-on in their criminal network participation. Like the Rockers in the previous Hells Angels network, the smaller gang members were the highest in various forms of direct connectivity, and, consequently, the most visible/vulnerable. The arrests made across these three investigations provide further support for the strategic positioning of some and the vulnerability of others: 36% of Bo-Gars members were arrested; 92% of the smaller gang members were arrested; and 62% of the nongang members were arrested. Recall, once again, that the investigations were organized to target the most reputed of the gang members in this network—the Bo-Gars members.

V. Conclusion

How can these results be interpreted? Once again, the highest-ranked brokers in the network were not gang members. As a group, however, brokerage capital was more heavily concentrated amongst the more reputed and higher status gang members. If we follow the previous case studies in this book, one interpretation would stress that most gang members are simply facilitators and aids to the core traffickers in the network. Another interpretation would maintain that gang members are key participants in this network because they have the capacity to govern the network from afar. This latter explanation provides some support for the basic hierarchy or pecking order structure that is often put forward as the evident representation of street gang settings in popular culture and law-enforcement circles. However, the pecking order that was in place revolved around individual members.

The more reputed and experienced gang members (the Bo-Gars members) did not act as a gang per se. They did not communicate extensively with each other. Only members from smaller, less reputed gangs were more cliquish in their networking throughout these drug-distribution operations—a pattern which is likely more typical of smaller, younger gangs. This suggests that gang members, like more general organized crime and criminal-market participants, improve their entrepreneurial capacities and personal network organization as the gang grows in size and reputation, and as their careers permit them to expand toward a wider and more dispersed range of opportunities. In terms of organizational structure, gangs are not that different from other criminal enterprise participants since adaptive capacities are required for the common volatile environment.

Chapter 10
Summary and Extensions

That criminal networks and organized criminal groups are less centralized and less tightly structured than often believed in popular venues and by law-enforcement and prosecutorial actors has been a critical source of argument in past research. The traditional divide between criminal justice and scholarly actors has been one where the former generally advocates the common stereotypes associated with a bureaucratic image of organized crime, while the latter dispels such an image through empirical research leading to representations of criminal groups as volatile resource-sharing ventures.

In diverse ways, the analytical chapters in this book provide further support for much of what past research has found and warned us against. Overall, clearly identifiable central participants emerged across all criminal networks, however, the designation and form of key player status did not coincide with that expected and put forward by the law-enforcement officials who gathered the data used for these case studies. The case studies across the book's chapters illustrated a variety of ways in which the coordination and general organization of criminal ventures hinge more on features that extend from a network's flexibility than on rigid organizational forms that keep participants or members in a systematic and well-ordered working structure. Chapter 3 illustrated that, in many cases, criminal networks are closer to simple co-offending or partnership ventures than they are to the sophisticated criminal organizations that they are often made out to be. Chapter 4 demonstrated that, although security may be the principal concern for most criminal-network participants, shaping a network in favor of increased participant insulation is typically a luxury which is more likely found in less profit-oriented networks, such as those executing terrorist operations. Chapter 5 expanded on this security–efficiency trade-off by examining how security is upheld in a typical profit-oriented criminal network. As demonstrated in previous research, the answer is in the periphery and the multitude of facilitators from legitimate spheres of society who share their

C. Morselli, *Inside Criminal Networks*, DOI 10.1007/978-0-387-09526-4_10,

status and resources as legitimate entrepreneurs not only to insulate the more evident criminal trade participants in the network, but also to generate a good portion of the network.

Chapter 6 carried the book's analytical flow in another direction by focusing on change and adaptation in a criminal network as it was disrupted by police interventions. This chapter confirmed previous arguments maintaining shifts in leadership and central participants, an overall tendency for criminal networks to decentralize when confronted with external controls, and a capacity for network participants to take matters into their own hands by adjusting to the challenges of criminal venturing without the governance of a dominant actor. In many ways, Chapter 6 is the heart of the flexible order thesis that is pursued across this book—it illustrated how structure and working order is nevertheless maintained in a decentralized setting that is in constant flux.

The remaining chapters in the book centered on what can be described as the key position in a flexible-order setting—that of the broker. Chapter 7 demonstrated the extent to which criminal networks are shaped around the coordinating capacity of brokers in the network. This was fundamental not only for the network itself, but for the overall crime-commission process that was in play. Chapters 8 and 9 pursued the brokerage focus by assessing how such an important position fits in the traditional hierarchical-like understanding of organized crime and criminal enterprise. The case studies carry us into street-gang settings and within the extended confines of a formal organization that over the years has gained the label of a criminal organization—the Hells Angels in Quebec. These last two case studies illustrated that the most reputed and elite members in these settings were not the hands-on and domineering participants that they are too often assumed to be. When compared to lower-level or less-reputed members of their organizational/gang confines, they were, above all, participants with lower direct connectivity and higher brokerage capital. Note, however, that whereas many may conclude that high-status gang members are more likely positioned as brokers and this should now be the focus of tactical and strategic law-enforcement interventions in such settings, we must recall that these case studies also pointed out that the top brokers in the street gang and Hells Angels case studies were not members of any gang or organization—they were independent entrepreneurs who coparticipated with members from such fixed groups. The most important point that must be retained after all this is that we have to stop framing these criminal activities as formal organizational boundaries and take them for what they are—fluid structures of criminal coparticipation in which, over time, certain participants come to position themselves better and more securely than others. Positioning in criminal networks is about fitting in loosely and not about taking control.

At the same time, the flexible order thesis, as presented in Chapter 1, does require some modification in light of some of the findings from the case studies. Before starting the research leading up to this book, I expected networks and hierarchical structures to be independent organizational forms. The case studies led me to see that it is not an either/or scenario. The network is a transcending organizational form that may include hierarchical subsets in as much as it may integrate freelance participants and ephemeral groups. The issue at hand, however, concerns network contexts in which hierarchical partitions may emerge as dominant structuring elements. While none of the networks studied in this book represented such a scenario, the latter two case studies did reveal that the presence of members from fixed groups within the network led primarily to a division in key participant status. Whereas the same participants ranked highest in both direct and brokerage-like centralities in the Ciel, Caviar, Siren, and Togo networks, distinctions between these forms of key player positions emerged in the street gang and Hells Angels case studies. In these latter networks, formal ranks or recognition in the street gang landscape or biker club may have influenced how members from these fixed groups were positioned to participate in the criminal activities. While we are far from the claim that gangs or bikers have the power to establish and enforce order in the scope of such criminal activities, we may nevertheless conclude that, at least amongst themselves, members from such fixed groups position themselves so that the pecking order in their hierarchy mirrors the more privileged positions in the general scheme of flexible order that they and their non-gang coparticipants find themselves as they converge in a variety of criminal operations.

I. Extensions of the Criminal Network Perspective

Social network analysis is not simply a set of methodological tools—it is a perspective, a way of seeing and approaching specific problems. The contributions from research within this perspective are wide and their impact in criminology has become increasingly pronounced with novel applications in areas within and beyond the scope of this book. Research on deviant peers has benefited from the necessary nuance that it is not simply the association with a deviant peer that matters, but how such associations are integrated within the wider scope of peers (Weerman and Smeenk 2005; Haynie 2001). Co-offending research has broadened the scope of investigation set forth by Reiss (1986) by employing a network perspective to search for nodal offenders (or frequent co-offenders) (Sarnecki 2001), estimate the size of the criminal population and offending patterns therein

(Frank and Carrington 2007), and assess the stability of co-offending over an extended life course (McGloin et al. 2008). Research on street gangs and intergroup violence has advanced our understanding of third-party dynamics by applying the basics of balance theory and a network framework (Papachristos 2007; Nakamura, Tita, and Krackhardt 2006; Gould 2003). Indeed, the applications of the ideas, methods, and general outlook of the social network field are already substantial in criminology. These last sections add to such research endeavors and propose a series of paths that extend from the case studies and criminal network perspective developed in this book.

Assessing Change Within Disrupted Networks

Further research is required to understand the adaptive capacities of criminal networks, at both individual and collective levels. This was a central theme in Cornish's (1994) script model in that "counter-countermoves" in the form of script modifications or offender replacement (Ekblom 1999) are a strong possibility. Although the responses of the Caviar network participants to law-enforcement targeting were illustrated in Chapter 6, conducting such analyses on a systematic basis is a more serious problem that extends from the use of law-enforcement data. Such material rarely permits an assessment of what actually happens after a key participant is removed or after a key feature of the crime-commission process is damaged. An investigation does not generally follow through on the remnants of a targeted operation to see if and how remaining participants react after their network is disrupted or a portion of coparticipants are arrested. While this became a central focus in the Project Caviar and Projects Siren/Togo case studies, such a concern is rarely addressed by law-enforcement agencies targeting these criminal networks and researchers interested in such matters. Similarly, the vast efforts to contain and dismantle the drug-distribution operations in either the street-gang setting in Montreal North or in the Hells Angels case study were also lacking such a strategic reappraisal of the law-enforcement disruption that took place.

The question, here, concerns not only who will survive in the criminal network, but also who is likely to succeed in taking over the key brokerage positions that are at the center of criminal network continuity. Also, whereas criminal network disruption is typically associated with participant removal, other research endeavors should also look into how the inclusion of participants (e.g., undercover law-enforcement agents, informants, agents provocateurs) affect a criminal network structure and the positioning of key players therein.

Content Analysis of Electronically Monitored Conversations

One aspect of the data sources compiled for this book that was analyzed minimally across the case studies is the content of conversations that were intercepted by law-enforcement investigators. As past researchers have illustrated, content analysis of this material is compatible with centrality and other network analyses. Such an approach adds to our understanding of the structure of the network and the form of order that is practiced therein. Natarajan (2006, 2000) used a hypertext software program to code the conversations between participants in the drug-importation networks that she analyzed. This allowed her to identify the roles that participants had within the network and designate the status of core participants. In both studies, Natarajan used a six-point scoring scheme that helped code a conversation between two participants for any possible status differentiation. Her coding scheme was consistent with the idea of symmetrical/asymmetrical relationships in that interactions were scrutinized for expressions of satisfaction, information seeking, order giving, and the use of status keywords such as "sir."

More recently, Varese (2006) analyzed the network of a Russian crime group as it attempted to implement some of its operations in Rome during the late 1990s. Each conversation that was intercepted by Italian police was analyzed to identify task allocation within the network. These tasks ranged from resource acquisition, investment in the economy, involvement in protection rackets, and maintaining internal order. Unlike Natarajan, Varese revealed the formidable challenge facing researchers who were intent on conducting similar analyses. Because most of the conversations were informal, an automatic coding scheme could not be designed to conduct an accurate content analysis. Instead, Varese argued in favor of a manual coding of all intercepted conversations.

Aside from Natarajan's (2000) first case study, which essentially collaborated what was expected from the prosecutor's depiction of the criminal network, the criminal networks under analysis in these past studies were found to be loosely structured or complying with a rudimentary division of labor. Natarajan (2006) described the network in her second case study as an egalitarian structure with little status differentiation. Varese (2006) reported a network that was divided along ethnic lines which failed to integrate over time. In both case studies, the content analysis of conversations was consistent with the network analysis of the criminal operations under examination.

Such research endeavors are indeed scant, but they nevertheless demonstrate the potential for incorporating new analytical procedures within the scope of law-enforcement data and criminal-network research designs.

Content analysis is indeed the next step toward enriching the various analyses conducted throughout this book.

Expanding on the Social Embeddedness of Network Participants

Another research angle that was minimally exploited in this book concerns the social embeddedness of the participants across the diverse networks. By restricting the focus of analysis to those relationships that were retrievable through telephone and physical surveillance, it was not possible to explore the details concerning the pre-existing and affective relationships between participants. Such analyses were beyond the scope of our study and would require a more complete analysis of individual arrest and intelligence records. Interviews with the participants themselves would clearly be the most suitable and straightforward approach for addressing this particular issue. Of course, gathering such data would also contribute more extensively to understanding how daily experiences, interactions, and chance occurrences contribute to the emergence of criminal networks in various settings. Such an approach would be rooted in previous ethnographic work by Whyte (1943/1993) and Ianni (1972), as well as more recent biographical research by Steffensmeier and Ulmer (2005) and Van Calster (2006). Indeed, there is much more to tell about criminal networks than the work presented in this book.

Criminal Network Position and Individual Traits[1]

In as much as the obvious focus of a network analysis would be the ensemble of contacts uniting individuals in a given setting, other aspects in such settings also require further understanding. There is an important volume of research on the relationships between individual-level differences in network positioning, social psychological traits, and achievement/performance outcomes. Whereas this book's case studies focused exclusively on "whole" network analyses and the place of key players therein, an additional branch of research should follow through on exploring the determinants and outcomes of various positions within criminal networks.

One important path centers on the brokerage position. Recall Boissevain's (1974) emphasis that one who takes on a brokerage role is ready to take part in the manipulation of others. Although manipulation has been excluded in research on such matters, some personality traits have been

[1] I thank Garry Robins for sharing his thoughts and past research on this issue.

consistent amongst brokerage-like players. Again, research from legitimate settings offer much insight on what may be expected in a criminal context. While emphasizing that personality data were not a substitute for sociometric data, Burt, Jannotta, and Mahoney (1998: 74–76) found that people with entrepreneurial networks (or greater brokerage capital) "claim the personalities of people who are the authors of their own world." They perceived themselves as "independent outsiders," "in search of authority (rather than security)," and who thrived on "advocacy and change (rather than stability)."

Within the criminal context, results from an inmate survey in southern Quebec found that offenders who had personal criminal networks that were greater in brokerage capital were not only those who had greater criminal earnings, but also those who reported lower self-control—low self-control was also a factor that accounted for higher criminal earnings (Morselli and Tremblay 2004). Carried into criminology in a provocative book by Gottfredson and Hirschi (1990), some of the behavioral components making up low self-control are consistent with Boissevain's rendition of the manipulative (or capitalizing) broker. This common ground should be incorporated in criminological research. McGloin et al. (2008), for example, found that juvenile offenders with lower self-control ratings were less stable in their co-offending patterns. Their interpretation of this finding followed Gottfredson and Hirschi's argument that individuals with low self-control were unable to forge lasting relationships, thus, making any criminal context typically unstable. But Gottfredson and Hirschi's view of the offender is one of an actor with limited competency and social skills. If we consider the brokerage angle, its association with the low self-control trait and the positive relationship with criminal earnings, such network positioning and personality traits are not consistent with the claim that offenders cannot forge lasting relationships. Co-offending contexts should not be framed in a similar manner as friendship or peer relationships—the underlying traits of low self-control and manipulative brokerage are not conducive to such settings. If, instead, we frame co-offending contexts as arenas in which collaboration is a necessity, but the ultimate driving force is individual competition and ambition, the network brokerage and low self-control components are not suggesting a lack of social skills, but a capacity to exploit and take advantage when necessary.

At the same time, not all is advantageous for the criminal broker. Kalish and Robins (2006) also found that individuals with greater brokerage capital were more likely to be individualists who sought autonomy in their lives, but they also found such people to have higher levels of neuroticism. They interpreted this result as a reflection of the stress associated with keeping friends separate, a mistrust of others, and the strife of managing a possible divide-and-rule personal network. Again, we are not far from the components of low self-control. At the same time, that more prominent criminal trade participants with greater brokerage capital may also experience greater neuroticism is compatible with criminological research that emphasizes strain as a source of crime (Agnew 2001, 1997, 1992, 1984) and despair as a possible extension (Hagan 1997).

Criminal Networks and the Criminal Career

One of the findings from the latter case studies in this book is that a participant's age matters within the criminal network scheme. In the Hells Angels case study, older participants were positioned as brokers, while younger participants bared the risks associated with direct connectivity within a criminal network. At the same time, age was also correlated with formal ranks, with older participants holding higher membership status within the Hells Angels. In the street-gang setting, an age influence was also indicated with members of older, more reputed gangs positioned as brokers within the network, in contrast to the more visible and clustered members of younger, less reputed gangs.

For quite some time, criminal career research has been divided into two separate strands of approaches and findings. Research from the mainstream strand, which recently (and curiously) blended life course approaches into the criminal career framework (Piquero, Farrington, and Blumstein 2003, 2007),[2] generally include large portions of one-time offenders in their samples and only a small portion of individuals who experienced long-term careers. While such research may provide us with a valuable representation of the criminal population, analyses are typically kept at the aggregate level and explorations of the more intimate details of doing crime are rarely considered. The alternative strand has relied primarily on qualitative studies of select careers in crime, in which the study's protagonist shares his/her experiences in a personal crime trajectory, while revealing the aptitudes and attitudes required for long-term and "successful" participation in crime. Unlike the mainstream strand, this approach centers on unique cases in the criminal career and how they stand out from the typical and majority of offenders. Steffensmeier and Ulmer (2005) are the most recent to contribute to this approach, with a follow-up to Steffensmeier's (1986) previous case study of a career fence. The divide in criminology is well pronounced with one

[2]Clearly, the inclusion of longitudinal data sets and methods are strong incentives for including life-course research in the criminal-career framework. Maybe one of the more important drawbacks of integrating this approach is that these same data sets began so far back that drug-dealing/trafficking and other market crimes are not always incorporated in the list of crimes.

side stressing the low-end offending of many and the other emphasizing the criminal capital of those who do get ahead through crime.

Some research has breached the divide between these two strands and has succeeded in testing the within-individual findings from case study research on criminal careers with similar aggregate data and methods as the mainstream strand. For example, researchers in the Netherlands have turned to sizeable data sets with a longitudinal outlook on criminal careers to identify a pattern in which older offenders help sustain the social opportunity structure for crime through their co-offending with younger offenders (Kleemans and dePoot 2008). Such a finding is consistent with results from an inmate survey that pointed to the enhancing impact of mentor relationships on the criminal career (Morselli, Tremblay, and McCarthy 2006).

The criminal network perspective can benefit from insights and findings from both these criminal career strands while also contributing with its specific focus on the social organization of offending groups. The aim would be toward assessing networking patterns at different phases of the career. Judging by the findings from this book's case studies, recent results from criminal career research, and a general fact from co-offending research that has consistently illustrated that co-offending declines as offenders get older, we would expect criminal networking patterns to shift throughout the criminal career in the following manner: a high degree of direct connectivity at early phases (thus increasing the likelihood of being arrested with co-offenders); a mix of high volume of direct contacts and strategic re-organization at mid-phases; and strategic brokerage-like networking at a later, more advanced phase (thus decreasing the likelihood of co-offender presence).

Criminal Networks Beyond the Individual

The criminal network perspective is not limited to interactions between individuals. A network framework may be applied to a number of contexts, as past research on interactions between gangs has illustrated. With so much attention directed toward globalization and its effects on crime patterns across the world, one of the more promising applications of this perspective can be foreseen within interactions between countries.

Criminal market research would benefit from the relational focus of a network framework. Past research on global drug markets, for example, has assessed source–destination patterns between countries (Farrell, Mansur, and Tullis 1996; Reuter and Kleiman 1986). Compiling information on a series of drug-trafficking routes in an asymmetrical and valued matrix would help reveal patterns not only about source and destination countries, but also about those countries that emerge as consistent transit (or broker) points along illegal drug and other illegal commodity routes. As with any representation of a criminal network, the matrix and eventual sociogram would help flesh out patterns and various forms of central nodes. Data on the number and weight of drug seizures may be used to bring valued measures to the matrix. Additional data on legitimate trade routes between countries can also be incorporated to assess the extent to which criminal networks resemble and overlap with legitimate networks on a global scale. Indeed, there is an impressive set of past research on international trade, global transport systems, communication patterns, and the world system perspective to extend from (Smith and Timberlake 2001; Kick and Davis 2001; Rossem 1996; Smith and White 1992; Snyder and Kick 1979; Wallerstein 1974). Symbiosis and the features of organizing crime may also be relevant within this criminal–legitimate trade overlap, creating a possible new set of intervention points for preventive measures. Finally, incorporating the network perspective would clearly contribute to criminal market research by revealing how illegal drug prices at various levels vary in accordance with a country's position within the wider network of illegal drug circulation.

The Master Network

An important drawback of the research method and data sources used throughout the book is that they do not lead to a representation and understanding of a possible master criminal network that is perpetually in place. To clarify, this master network is not the stuff of criminal conspiracies; it is, instead, a matter of criminal opportunity and necessary collaboration. All of the networks studied here drew from the same pool of potential co-offenders that extended from the Montreal region. These networks were not independent from one another—neither ethnic segregation nor exclusive affiliation to specific criminal groups emerged as important features in these networks. Indeed, some level of overlap should be expected between participants from the different case studies, particularly since most operated within similar time frames. Furthermore, and because most of the cases analyzed here dealt with the importation or exportation of an illegal commodity, the master network that spans these case studies is not likely to be restricted to the Montreal region. Considering that many of the participants in the criminal networks resided in several different countries across Europe, Africa, Asia, and South America, a next (and more ambitious) step for criminal network analysis would be to model the small-world dynamics that structure the vast array of transnational criminal activities. Doing so would test the flexible order thesis in a more extended context, while adding some credence for or against repeated and sensationalist claims of a global crime conspiracy. Arriving at such a representation would also contribute to other efforts at estimating criminal populations (Bouchard and Tremblay 2005). Aside from simulation approaches, the most promising method for arriving at a more complete representation would involve snowball (or respondent-driven) sampling procedures that have been developed extensively for hidden population research (see Salganik and Heckathorn 2004; Heckathorn and Jeffri 2001).

With this said, I repeat that the strategic cases that were analyzed here and in which the properties of criminal networks were drawn may be seen as snapshots of this master network. The task that I laid out for myself in this book was to assess the structure and inner workings of such organizational forms that for some fixed period of time came under the observational scope of law-enforcement targeting. In the end, by limiting my observational scope as such, I am still subjected to one of many criticisms put forward by Felson (2003) in regard to the use of social network analysis in criminology: *a social network for crime, as important as it might be, generates a serious problem, since a network has no clear boundaries and is difficult to measure, analyze, or use to predict what happens on the ground. Somebody has got to specify the facts about a delinquency network and show that it has an ongoing structure producing criminal cooperation* (p. 155). For the moment, I am not that somebody. The criminal networks studied here may be best described as segments of that wider network that is consistently in place and from which smaller and more ephemeral sub-networks emerge. Unfortunately, I did not have the empirical material to study the master (and not master's) network. This task will have to wait for a later time.

Appendix

The Ciel Network

	1	2	3	4	5	6	7	8	9	10	11	12	13	14	15	16	17	18	19	20	21	22	23	24	25
1	0	17	1	4	0	17	0	0	0	0	0	0	1	0	3	1	0	1	0	1	0	1	0	0	0
2	21	0	0	0	0	0	5	1	0	0	15	1	0	2	0	0	1	0	2	0	0	0	0	1	1
3	0	0	0	0	0	0	0	0	0	0	0	0	0	0	0	0	0	0	0	0	0	0	0	0	0
4	0	0	1	0	0	0	0	0	0	0	0	0	0	0	0	0	0	0	0	0	0	0	0	0	0
5	0	0	0	1	0	0	0	0	0	0	0	0	0	0	0	0	0	0	0	0	0	0	0	0	0
6	0	0	0	0	0	0	0	0	0	0	0	0	0	0	0	0	0	0	0	0	0	0	0	0	0
7	0	1	0	0	0	0	0	0	0	0	0	0	0	0	0	0	0	0	0	0	0	0	0	0	0
8	0	0	0	0	0	0	0	0	1	0	0	0	1	0	0	0	0	0	0	0	0	0	0	0	0
9	0	0	0	0	0	0	0	0	0	0	0	0	8	0	0	0	0	0	0	0	0	0	0	0	0
10	0	22	0	0	0	0	1	1	0	0	1	0	0	0	0	0	0	0	0	0	0	0	0	0	0
11	0	10	0	0	0	0	0	0	0	0	0	0	0	0	0	0	0	0	0	0	0	0	0	10	0
12	0	0	0	0	0	0	0	0	0	0	0	0	0	0	0	0	0	0	0	0	0	0	0	0	0
13	0	0	0	0	0	0	1	2	9	0	0	0	0	0	0	0	0	0	0	0	0	1	0	0	0
14	0	2	0	0	0	0	0	0	0	0	0	0	0	0	0	0	0	0	0	0	0	0	0	0	0
15	2	0	0	1	0	0	0	0	0	0	0	0	0	0	0	0	0	0	0	0	0	0	0	0	0
16	1	0	0	0	0	0	0	0	0	0	0	0	0	0	0	0	0	0	0	0	0	0	0	0	0
17	0	0	0	0	0	0	1	0	0	0	0	0	0	0	0	0	0	0	0	0	0	0	0	0	0
18	0	0	0	0	0	0	0	0	0	0	0	0	0	0	0	0	0	0	0	0	0	0	0	0	0
19	0	0	0	0	0	0	0	0	0	0	0	0	0	0	0	0	0	0	0	0	0	0	0	0	0
20	0	0	0	0	0	0	0	0	0	0	0	0	0	0	0	0	0	0	0	0	0	0	0	0	0
21	0	1	0	0	0	0	0	0	0	0	0	0	0	0	0	0	0	0	0	0	0	0	0	0	0
22	0	0	0	0	0	0	0	0	0	0	0	0	1	0	0	0	0	0	0	0	0	0	0	0	0
23	0	0	0	0	0	0	0	0	0	0	20	0	0	0	0	0	0	0	0	0	0	0	0	0	0
24	0	0	0	0	0	0	0	0	0	0	1	0	0	0	0	0	0	0	0	0	0	0	0	0	0
25	0	0	0	0	0	0	0	0	0	0	0	0	0	0	0	0	0	0	0	0	0	0	0	0	0

The Caviar Network (Complete/Part 1: Upper-Left Quarter)

	1	2	83	3	85	82	88	89	4	5	90	86	6	7	8	9	10	11	12	76	84	13	14	15	16	17	18	77	19	78	20	87	21	22	79	23	24	25	26	27	28	29	30	31	32	33	80	34	35	36	81	37	38	39	91
1	0	52	81	337	142	73	9	30	10	49	7	35	25	1	46	39	6	12	11	88	4	11	3	11	0	0	0	28	21	0	7	68	0	0	0	0	0	0	0	0	3	0	0	0	0	0	0	4	1	0	7	2	0	0	0
2	29	0	0	3	0	0	0	0	0	1	0	0	0	0	0	0	0	0	0	0	0	0	0	0	0	0	0	0	0	0	0	0	0	0	0	0	0	0	0	0	0	0	0	0	0	0	0	2	1	0	0	0	0	0	0
83	6	0	0	1	0	0	0	0	0	0	0	25	0	0	0	0	0	0	0	0	5	0	0	0	0	0	0	0	0	0	0	0	0	0	0	0	0	0	0	0	0	0	0	0	0	0	0	0	0	0	0	0	0	0	0
3	152	1	5	0	29	2	2	2	0	0	1	4	0	1	21	16	0	0	21	25	0	0	1	0	0	0	0	3	0	0	0	18	0	0	0	0	0	0	0	0	0	0	0	11	0	0	0	0	0	0	0	0	0	0	0
85	33	0	0	18	0	0	12	0	0	0	0	0	0	0	2	0	0	0	0	9	0	0	0	0	0	0	0	0	0	0	0	0	0	0	0	0	0	0	0	0	0	0	0	0	0	0	0	0	0	0	0	0	0	0	0
82	18	0	0	2	0	0	0	0	0	0	0	0	0	0	0	0	0	0	0	0	0	0	0	0	0	0	0	0	0	0	0	4	0	0	0	0	0	0	0	0	0	0	0	0	0	0	0	0	0	8	0	0	0	0	0
88	2	0	0	1	10	0	0	0	0	0	0	0	2	0	0	0	0	0	0	0	0	0	0	0	0	0	0	0	0	0	0	0	1	0	0	0	0	0	0	0	0	0	0	0	0	0	0	0	0	0	0	0	0	0	0
89	2	0	0	1	0	0	0	0	0	0	0	0	0	7	0	0	0	0	2	0	0	0	1	0	0	0	0	0	0	0	0	0	0	0	0	0	0	0	0	0	0	0	0	0	0	0	0	0	0	0	0	0	0	0	0
4	1	0	0	0	0	0	0	0	0	1	0	0	0	0	0	0	0	0	0	0	0	0	0	0	0	0	0	0	0	0	0	0	0	0	0	0	0	0	0	0	0	0	0	0	0	0	0	0	0	0	0	0	0	0	0
5	65	0	0	3	0	0	0	0	0	0	0	0	0	0	2	0	0	0	0	0	0	0	0	0	0	0	0	0	0	0	0	0	0	0	0	0	0	0	0	0	0	0	0	0	0	0	0	0	0	0	0	0	0	0	0
90	1	0	0	1	0	0	0	0	0	0	0	0	0	0	0	0	0	0	0	0	0	0	0	0	0	0	0	0	0	0	0	0	0	0	0	0	0	0	0	0	0	0	0	0	0	0	0	0	0	0	0	0	0	0	0
86	5	0	19	1	0	0	0	0	0	0	0	0	0	0	0	0	0	0	0	0	0	0	0	0	0	0	0	0	0	0	0	0	0	0	0	0	0	0	0	0	0	0	0	0	0	0	0	0	0	0	0	0	0	0	0
6	41	0	0	1	0	0	0	0	0	0	0	0	0	0	0	0	0	0	0	0	0	0	0	0	0	0	0	0	0	0	0	0	0	0	0	0	0	0	0	0	0	0	0	0	0	0	0	0	0	0	0	0	0	0	0
7	0	0	0	0	0	0	0	7	0	0	0	0	0	0	0	0	0	0	0	0	0	0	0	0	0	0	0	0	0	0	0	0	0	0	0	0	0	0	0	0	0	0	0	0	0	0	0	0	0	0	0	0	0	0	0
8	10	0	0	1	0	0	0	0	0	1	0	0	0	0	0	7	0	0	0	0	0	0	0	0	0	0	0	0	0	1	0	0	0	0	0	0	0	0	0	0	0	0	0	0	0	0	0	0	1	0	0	0	0	0	0
9	0	0	0	0	0	0	0	0	0	0	0	0	0	0	2	0	0	0	0	0	0	0	0	0	0	0	0	0	0	0	0	0	0	0	0	0	0	0	0	0	0	0	0	0	0	0	0	0	0	0	0	0	0	0	0
10	0	0	0	0	0	0	0	0	0	0	0	0	0	0	0	0	0	0	0	0	0	0	0	0	0	0	0	0	0	0	0	0	0	0	0	0	0	0	0	0	0	0	0	0	0	0	0	0	0	0	0	0	0	0	0
11	118	0	0	1	0	0	0	0	0	0	0	0	0	0	0	0	0	0	0	0	0	0	0	0	0	0	0	0	0	0	17	0	0	0	0	0	0	0	0	0	0	0	0	0	0	0	0	0	0	0	0	0	0	0	0
12	1	0	0	25	0	0	0	0	0	0	0	0	0	0	0	2	0	0	0	1	0	4	73	0	51	12	17	0	0	0	0	0	0	6	14	1	1	2	2	2	0	3	1	3	0	0	0	0	0	0	0	0	0	0	0
76	66	0	1	6	7	0	0	1	0	0	0	0	0	0	0	0	0	0	1	0	0	0	0	0	0	0	0	0	5	12	0	3	0	0	4	0	0	0	0	0	0	0	0	0	0	0	0	0	0	0	0	0	0	0	0
84	9	0	1	0	0	0	0	0	0	0	0	0	0	0	0	0	0	0	0	0	0	0	0	0	0	0	0	0	0	0	0	0	0	0	0	0	0	0	0	0	0	0	0	0	0	0	0	0	0	0	0	0	0	0	0
13	0	0	0	0	0	0	0	0	0	0	0	0	0	0	0	2	0	0	0	0	0	0	0	0	0	0	0	0	0	0	0	0	0	0	0	0	0	0	0	0	0	0	0	0	0	0	0	0	0	0	0	0	0	0	0
14	1	0	0	0	0	0	0	0	0	0	0	0	0	0	1	0	0	0	64	0	0	0	0	0	0	0	0	0	0	0	0	0	0	0	0	0	1	0	0	0	0	0	0	0	0	0	0	0	0	0	0	0	0	0	0
15	2	0	0	1	0	0	0	0	0	0	0	0	0	0	0	0	0	0	0	2	0	0	0	0	0	0	0	0	0	0	0	0	0	0	0	0	0	0	0	0	0	0	0	0	0	0	0	0	0	0	0	0	0	0	0
16	0	0	0	0	0	0	0	0	0	0	0	0	0	0	0	0	0	0	93	0	0	0	0	0	0	0	0	0	0	0	0	0	0	0	0	0	0	0	0	0	0	0	0	0	0	0	0	0	0	0	0	0	0	0	0
17	0	0	0	0	0	0	0	0	0	0	0	0	0	0	0	0	0	0	37	0	0	0	0	0	0	0	0	0	0	0	0	0	0	0	0	0	0	0	0	0	0	0	1	0	0	0	0	0	0	0	0	0	0	0	0
18	0	0	0	0	0	0	0	0	0	0	0	0	0	0	0	0	0	0	44	0	0	0	0	0	0	0	0	0	0	0	0	0	0	0	0	0	0	0	0	0	0	0	0	0	0	0	0	0	0	0	0	0	0	0	0
77	15	0	0	0	0	0	0	0	0	0	0	0	0	0	0	0	0	0	0	0	0	0	0	0	0	0	0	0	1	0	0	0	0	0	0	0	0	0	0	0	0	0	0	0	0	0	0	0	0	0	0	0	0	0	0
19	96	0	0	1	0	0	0	0	0	0	0	0	0	0	0	0	0	0	0	1	0	0	0	0	0	0	0	0	0	0	0	0	0	0	0	0	0	0	0	0	0	0	0	0	0	0	0	0	0	0	0	0	0	0	0
78	0	0	0	0	0	0	0	0	0	0	0	0	0	0	0	0	0	0	0	27	0	0	0	0	0	0	0	0	0	0	0	0	0	0	3	0	0	0	0	0	0	0	0	0	0	0	0	0	0	0	0	0	0	0	0
20	15	0	0	0	0	0	0	0	0	0	0	1	0	0	0	0	0	0	0	0	0	0	0	0	0	0	0	0	0	0	0	0	0	0	0	0	0	0	0	0	0	0	0	0	0	0	0	0	0	0	0	0	0	0	0
87	48	0	0	28	0	18	0	0	0	0	0	0	0	0	0	1	0	0	0	2	0	0	0	0	0	0	0	0	0	0	0	0	9	0	0	0	0	0	0	1	0	0	0	0	0	0	0	0	0	0	21	29	2	9	3
21	0	0	0	0	0	0	0	0	0	0	0	0	0	0	0	0	0	0	0	0	0	0	0	0	0	0	0	0	0	0	0	0	0	0	0	0	0	0	0	0	0	0	0	0	0	0	0	0	0	0	0	0	0	0	0
22	0	0	0	0	0	0	0	0	0	0	0	0	0	0	0	0	0	0	4	0	0	0	0	0	0	0	0	0	0	0	0	0	0	0	0	0	0	0	0	0	0	0	0	0	0	0	0	0	0	0	0	0	0	0	0
79	0	0	0	0	0	0	0	0	0	0	0	0	0	0	0	0	0	0	2	0	0	0	0	0	0	0	0	0	0	5	0	0	0	0	0	0	0	0	0	0	0	0	0	0	0	0	0	0	0	0	0	0	0	0	0
23	0	0	0	0	0	0	0	0	0	0	0	0	0	0	0	0	0	0	6	0	0	0	0	0	0	0	0	0	0	0	0	0	0	0	0	0	0	0	0	0	0	0	0	0	0	0	0	0	0	0	0	0	0	0	0
24	0	0	0	0	0	0	0	0	0	0	0	0	0	0	0	0	0	0	0	0	0	0	1	0	0	0	0	0	0	0	0	0	0	0	0	0	0	0	0	0	0	0	0	0	0	0	0	0	0	0	0	0	0	0	0
25	0	0	0	0	0	0	0	0	0	0	0	0	0	0	0	0	0	0	4	0	0	0	0	0	0	0	0	0	0	0	0	0	0	0	0	0	0	0	0	0	0	0	0	0	0	0	0	0	0	0	0	0	0	0	0
26	0	0	0	0	0	0	0	0	0	0	0	0	0	0	0	0	0	0	0	0	0	0	0	0	0	0	0	0	0	0	0	0	0	0	0	0	0	0	0	0	0	0	0	0	0	0	0	0	0	0	0	0	0	0	0
27	0	0	0	0	0	0	0	0	0	0	0	0	0	0	0	0	0	0	4	0	0	0	0	0	0	0	0	0	0	0	0	0	0	0	0	0	0	0	0	0	0	0	0	0	0	0	0	0	0	0	0	0	0	0	0
28	0	0	0	0	0	0	0	0	0	0	0	0	0	0	0	0	0	0	0	0	0	0	0	0	0	0	0	0	0	0	0	0	0	0	0	0	0	0	0	0	0	0	0	0	0	0	0	0	0	0	0	0	0	0	0
29	0	0	0	0	0	0	0	0	0	0	0	0	0	0	0	0	0	0	4	0	0	0	0	0	0	0	0	0	0	0	0	0	0	0	0	0	0	0	0	0	0	0	0	0	0	0	0	0	0	0	0	0	0	0	0
30	0	0	0	0	0	0	0	0	0	0	0	0	0	0	0	0	0	0	0	0	0	0	0	0	0	0	0	0	0	0	0	0	0	0	0	0	0	0	0	0	0	0	0	0	0	0	0	0	0	0	0	0	0	0	0
31	0	0	0	0	0	0	0	1	0	0	0	0	0	0	0	0	0	0	0	0	0	1	0	0	0	0	0	0	0	0	0	0	0	0	0	0	0	0	0	0	0	0	0	0	0	0	0	0	0	0	0	0	0	0	0
32	1	0	0	0	0	0	0	0	0	0	0	0	0	0	0	0	0	0	1	0	0	0	0	0	0	0	0	0	0	0	0	0	0	0	0	0	0	0	0	0	0	0	0	0	0	0	0	0	0	0	0	0	0	0	0
33	0	0	0	0	0	0	0	0	0	0	0	0	0	0	0	0	0	0	1	0	0	0	0	0	0	0	0	0	0	0	0	0	0	0	0	0	0	0	0	0	0	0	0	0	0	0	0	0	0	0	0	0	0	0	0
80	0	0	0	0	0	0	0	0	0	0	0	0	0	0	0	0	0	0	1	0	0	0	0	0	0	0	0	0	0	0	0	0	0	0	0	0	0	0	0	0	0	0	0	0	0	0	0	0	0	0	0	0	0	0	0
34	1	1	0	0	0	0	0	0	0	0	0	0	0	0	0	0	0	0	0	0	0	0	0	0	0	0	0	0	0	0	0	0	0	0	0	0	0	0	0	0	0	0	0	0	0	0	0	0	0	0	0	0	0	0	0
35	0	0	0	0	0	0	0	0	0	0	0	0	0	0	0	0	0	0	0	0	0	0	0	0	0	0	0	0	0	0	0	0	0	0	0	0	0	0	0	0	0	0	0	0	0	0	0	0	0	0	0	0	0	0	0
36	0	0	0	0	0	0	0	0	0	0	0	0	0	0	0	0	0	0	0	0	0	0	0	0	0	0	0	0	0	0	0	0	0	0	0	0	0	0	0	0	0	0	0	0	0	0	0	0	0	0	0	0	0	0	0
81	0	0	0	0	0	0	0	0	0	0	0	0	0	0	0	0	0	0	0	0	0	0	0	0	0	0	0	0	0	0	0	2	0	0	0	0	0	0	0	0	0	0	0	0	0	0	0	0	0	0	0	0	0	0	0
37	1	0	0	0	0	11	0	0	0	0	0	0	0	0	0	0	0	0	0	0	0	0	0	0	0	0	0	0	0	0	0	12	0	0	0	0	0	0	0	4	0	0	0	0	0	0	0	0	0	0	0	0	0	0	0
38	0	0	0	0	0	0	0	0	0	0	0	0	0	0	0	0	0	0	0	0	0	0	0	0	0	0	0	0	0	0	0	1	0	0	0	0	0	0	0	0	0	0	0	0	0	0	0	0	0	0	0	0	0	0	0
39	0	0	0	0	0	0	0	0	0	0	0	0	0	0	0	0	0	0	0	0	0	0	0	0	0	0	0	0	0	0	0	0	0	0	0	0	0	0	0	0	0	0	0	0	0	0	0	0	0	0	0	0	0	0	0
91	0	0	0	0	0	0	0	0	0	0	0	0	0	0	0	0	0	0	0	0	0	0	0	0	0	0	0	0	0	0	0	0	0	0	0	0	0	0	0	0	0	0	0	0	0	0	0	0	0	0	0	0	0	0	0

The Caviar Network (Complete/Part 2: Upper-Right Quarter)

	40	41	42	92	93	94	43	95	44	45	96	46	97	47	98	48	49	50	51	99	52	53	54	55	56	100	57	101	58	59	102	103	104	60	105	61	62	106	107	108	63	109	110	64	65	66	67	68	69	70	71	72	73	74	75
1	0	2	0	0	0	0	0	0	0	0	1	0	1	2	1	2	3	2	1	1	4	0	2	9	1	0	2	0	0	0	0	1	0	0	1	0	1	1	0	0	0	0	0	6	0	0	1	0	0	4	0	0	0	0	0
2	0	0	0	0	0	0	0	0	0	0	0	0	0	0	0	0	0	0	0	0	0	0	0	0	0	0	0	0	0	0	0	0	0	0	0	0	0	0	0	0	0	0	0	0	0	0	0	0	0	0	0	0	0	0	0
83	0	0	0	0	0	0	0	0	0	0	0	0	0	0	0	0	0	0	0	0	3	0	0	0	0	0	0	0	0	0	0	0	0	0	0	0	0	0	4	0	0	0	0	0	0	0	0	0	0	0	0	0	0	0	0
3	0	0	0	0	0	0	0	0	0	0	0	0	0	0	0	0	0	0	0	1	0	0	0	0	0	0	0	0	0	0	0	0	0	0	0	0	0	0	0	0	0	0	0	0	0	0	0	0	0	0	0	0	0	3	0
85	0	0	0	0	0	0	0	0	0	0	0	0	0	0	0	0	0	0	0	0	0	0	0	0	0	0	0	0	0	0	0	0	0	0	0	0	0	0	0	0	0	0	0	0	0	0	0	0	0	0	0	0	0	0	0
82	0	0	0	0	0	0	0	0	0	0	15	25	0	0	0	0	0	0	0	0	0	0	0	0	0	0	0	0	0	0	0	0	0	0	0	0	0	0	0	0	0	0	0	0	0	0	0	0	0	0	0	0	0	0	0
88	0	0	0	0	0	0	0	0	0	0	0	0	0	0	0	0	0	0	0	0	0	0	0	0	0	0	0	0	0	0	0	0	0	0	0	0	0	0	0	0	0	0	0	0	0	0	0	0	0	0	0	0	0	0	0
89	0	0	0	0	0	0	0	0	0	0	0	0	0	0	0	0	0	0	0	0	0	0	0	0	0	0	0	0	0	0	0	0	0	0	0	0	0	0	0	0	0	1	1	0	0	0	0	0	0	0	0	0	0	0	0
4	0	0	0	0	0	0	0	0	0	0	0	0	0	0	0	0	0	0	0	0	0	0	0	0	0	0	0	0	0	0	0	0	0	0	0	0	0	0	0	0	0	0	0	0	0	0	0	0	0	0	0	0	0	0	0
5	0	0	0	0	0	0	0	0	0	0	0	0	0	0	0	0	0	0	0	0	0	0	0	0	0	0	0	0	0	0	0	0	0	0	0	0	0	0	0	0	0	0	0	0	0	0	0	0	0	0	0	0	0	0	0
90	0	0	0	0	0	0	0	0	0	0	0	0	0	0	0	0	0	0	0	0	0	0	0	0	0	0	0	0	0	0	0	0	0	0	0	0	0	0	0	0	0	0	0	0	0	0	0	0	0	0	0	0	0	0	0
86	0	0	0	0	0	0	0	0	0	0	0	0	0	0	0	0	0	0	0	0	0	0	0	0	0	0	0	0	0	0	0	0	0	0	0	0	0	0	3	0	0	0	0	0	0	0	0	0	0	0	0	0	0	0	0
6	0	0	0	0	0	0	0	0	0	0	0	0	0	0	0	0	0	0	0	0	0	0	0	0	0	0	0	0	0	0	0	0	0	0	0	0	0	0	0	0	0	0	0	0	0	0	0	0	0	0	0	0	0	0	0
7	0	0	0	0	0	0	0	0	0	0	0	0	0	0	0	0	0	0	0	0	0	0	0	0	0	0	0	0	0	0	0	0	0	0	0	0	0	0	0	0	0	0	0	0	0	0	0	0	0	0	0	0	0	0	0
8	0	0	0	0	0	0	0	0	0	0	0	0	0	0	0	0	0	0	0	0	0	0	0	0	0	0	0	0	0	0	0	0	0	0	0	0	0	0	0	0	0	0	0	0	0	0	0	0	0	0	0	0	0	0	0
9	0	0	0	0	0	0	0	0	0	0	0	0	0	0	0	0	0	0	0	0	0	0	0	0	0	0	0	0	0	0	0	0	0	0	0	0	0	0	0	0	0	0	0	0	0	0	0	0	0	0	0	0	0	0	0
10	0	0	0	0	0	0	0	0	0	0	0	0	0	0	0	0	0	0	0	0	0	0	0	0	0	0	0	0	0	0	0	0	0	0	0	0	0	0	0	0	0	0	0	0	0	0	0	0	0	0	0	0	0	0	0
11	0	0	0	0	0	0	0	0	0	0	0	0	0	0	0	0	0	0	0	0	0	0	0	0	0	0	0	0	0	0	0	0	0	0	0	0	0	0	0	0	0	0	0	0	0	0	0	0	0	0	0	0	0	0	0
12	0	0	0	0	0	0	0	0	0	0	0	0	0	0	0	0	0	0	0	0	0	0	0	0	0	0	0	0	0	0	0	0	0	0	0	0	0	0	0	0	0	0	0	0	0	2	0	0	0	0	0	0	0	0	0
76	0	0	0	0	0	0	0	0	0	0	0	0	0	0	0	0	0	0	0	0	0	0	0	0	0	0	0	0	0	5	0	0	0	0	0	4	0	0	0	0	0	0	0	0	0	0	0	0	0	0	1	2	0	0	0
84	0	0	0	0	0	0	0	0	0	0	0	0	0	0	0	0	0	0	0	0	2	0	0	0	0	0	0	0	0	0	0	0	0	0	0	0	0	0	0	0	0	0	0	0	0	0	0	0	0	0	0	0	0	0	0
13	0	2	0	0	0	0	0	0	0	0	0	0	0	0	0	0	0	0	0	0	0	0	0	0	0	0	0	0	0	0	0	0	0	0	0	0	0	0	0	0	0	0	0	0	0	0	0	0	0	0	0	0	0	0	0
14	0	0	0	0	0	0	0	0	0	0	0	0	0	0	0	0	0	0	0	0	0	0	0	0	0	0	0	0	24	0	0	0	0	0	0	0	0	0	0	0	0	0	0	0	0	0	0	0	0	0	0	0	0	0	0
15	0	0	0	0	0	0	0	0	0	0	0	0	0	0	0	0	0	0	0	0	0	0	0	0	0	0	0	0	0	0	0	0	0	0	0	0	0	0	0	0	0	0	0	0	0	0	0	0	0	0	0	0	0	0	0
16	0	0	0	0	0	0	0	0	0	0	0	0	0	0	0	0	0	0	0	0	0	0	0	0	0	0	0	0	1	0	0	0	0	0	0	0	0	0	0	0	0	0	0	0	0	0	0	0	0	0	0	0	0	0	0
17	0	0	0	0	0	0	0	0	0	0	0	0	0	0	0	0	0	0	0	0	0	0	0	0	0	0	0	0	0	0	0	0	0	0	0	0	0	0	0	0	0	0	0	0	0	0	0	0	0	0	0	0	0	0	0
18	0	0	0	0	0	0	0	0	0	0	0	0	0	0	0	0	0	0	0	0	0	0	0	0	0	0	0	0	0	0	0	0	0	0	0	0	0	0	0	0	0	0	0	0	0	0	0	0	0	0	0	0	0	0	0
77	0	0	0	0	0	0	0	0	0	0	0	0	0	0	0	0	0	0	0	0	0	0	0	0	0	0	0	0	0	0	0	0	0	0	0	0	0	0	0	0	0	0	0	0	0	0	0	0	0	0	0	0	0	0	0
19	0	0	0	0	0	0	0	0	0	0	0	0	0	0	0	0	0	0	0	0	0	0	0	0	0	0	0	0	0	0	0	0	0	0	0	0	0	0	0	0	0	0	0	0	0	0	0	0	0	0	0	0	0	0	0
78	0	0	0	0	0	0	0	0	0	0	0	0	0	0	0	0	0	0	0	0	0	0	0	0	0	0	0	0	0	0	0	0	0	0	0	0	0	0	0	0	0	0	0	0	0	0	0	0	0	0	0	0	0	0	0
20	0	0	0	0	0	0	0	0	0	0	0	0	0	0	0	0	0	0	0	0	0	0	0	0	0	0	0	0	0	0	0	0	0	0	0	0	0	0	0	0	0	0	0	0	0	0	0	0	0	0	0	0	0	0	0
87	1	5	0	0	0	0	0	0	0	0	0	0	0	0	0	0	0	0	0	0	0	0	0	0	0	0	0	0	0	0	0	0	0	0	5	0	0	0	0	0	0	0	0	0	0	0	0	0	0	0	0	0	2	0	0
21	0	0	0	0	0	0	0	0	0	0	0	0	0	0	0	0	0	0	0	0	0	0	0	0	0	0	0	0	0	0	0	0	0	0	0	0	0	0	0	0	0	0	0	0	0	0	0	0	0	0	0	0	0	0	0
22	0	0	0	0	0	0	0	0	0	0	0	0	0	0	0	0	0	0	0	0	0	0	0	0	0	0	0	0	0	0	0	0	0	0	0	0	0	0	0	0	0	0	0	0	0	0	0	0	0	0	0	0	0	0	0
79	0	0	0	0	0	0	0	0	0	0	0	0	0	0	0	0	0	0	0	0	0	0	0	0	0	0	0	1	0	1	1	0	0	0	0	0	0	0	0	0	0	0	0	0	0	0	0	0	0	0	0	0	0	0	0
23	0	0	0	0	0	0	0	0	0	0	0	0	0	0	0	0	0	0	0	0	0	0	0	0	0	0	0	0	0	0	0	0	0	0	0	0	0	0	0	0	0	0	0	0	0	0	0	0	0	0	0	0	0	0	0
24	0	0	0	0	0	0	0	0	0	0	0	0	0	0	0	0	0	0	0	0	0	0	0	0	0	0	0	0	0	0	0	0	0	0	0	0	0	0	0	0	0	0	0	0	0	0	0	0	0	0	0	0	0	0	0
25	0	0	0	0	0	0	0	0	0	0	0	0	0	0	0	0	0	0	0	0	0	0	0	0	0	0	0	0	0	0	0	0	0	0	0	0	0	0	0	0	0	0	0	0	0	0	0	0	0	0	0	0	0	0	0
26	0	0	0	0	0	0	0	0	0	0	0	0	0	0	0	0	0	0	0	0	0	0	0	0	0	0	0	0	0	0	0	0	0	0	0	0	0	0	0	0	0	0	0	0	0	0	0	0	0	0	0	0	0	0	0
27	0	0	0	0	0	0	0	0	0	0	0	0	0	0	0	0	0	0	0	0	0	0	0	0	0	0	0	0	0	0	0	0	0	0	0	0	0	0	0	0	0	0	0	0	0	0	0	0	0	0	0	0	0	0	0
28	0	0	0	0	0	0	0	0	0	0	0	0	0	0	0	0	0	0	0	0	0	0	0	0	0	0	0	0	0	0	0	0	0	0	0	0	0	0	0	0	0	0	0	0	0	0	0	0	0	0	0	0	0	0	0
29	0	0	0	0	0	0	0	0	0	0	0	0	0	0	0	0	0	0	0	0	0	0	0	0	0	0	0	0	0	0	0	0	0	0	0	0	0	0	0	0	0	0	0	0	0	0	0	0	0	0	0	0	0	0	0
30	0	0	0	0	0	0	0	0	0	0	0	0	0	0	0	0	0	0	0	0	0	0	0	0	0	0	0	0	0	0	0	0	0	0	0	0	0	0	0	0	0	0	0	0	0	0	0	0	0	0	0	0	0	0	0
31	0	0	0	0	0	0	0	0	0	0	0	0	0	0	0	0	0	0	0	0	0	0	0	0	0	0	0	0	0	0	0	0	0	0	0	0	0	0	0	0	0	0	0	0	0	0	0	0	0	0	0	0	0	0	0
32	0	0	0	0	0	0	0	0	0	0	0	0	0	0	0	0	0	0	0	0	0	0	0	0	0	0	0	0	0	0	0	0	0	0	0	0	0	0	0	0	0	0	0	0	0	0	0	0	0	0	0	0	0	0	0
33	0	0	0	0	0	0	0	0	0	0	0	0	0	0	0	0	0	0	0	0	0	0	0	0	0	0	0	0	0	0	0	0	0	0	0	0	0	0	0	0	0	0	0	0	0	0	0	0	0	0	0	0	0	0	0
80	0	0	0	0	0	0	0	0	0	0	0	0	0	0	0	0	0	0	0	0	0	0	0	0	0	0	0	0	0	0	0	0	0	0	0	0	0	0	0	0	0	0	0	0	0	0	0	0	0	0	0	0	0	0	0
34	0	0	0	0	0	0	0	0	0	0	0	0	0	0	0	0	0	0	0	0	0	0	0	0	0	0	0	0	0	0	0	0	0	0	0	0	0	0	0	0	0	0	0	0	0	0	0	0	0	0	0	0	0	0	0
35	0	0	0	0	0	0	0	0	0	0	0	0	0	0	0	0	0	0	0	0	0	0	0	0	0	0	0	0	0	0	0	0	0	0	0	0	0	0	0	0	0	0	0	0	0	0	0	0	0	0	0	0	0	0	0
36	0	0	0	0	0	0	0	0	0	0	0	0	0	0	0	0	0	0	0	0	0	0	0	0	0	0	0	0	0	0	0	0	0	0	0	0	0	0	0	0	0	0	0	0	0	0	0	0	0	0	0	0	0	0	0
81	0	0	0	0	0	0	0	0	0	0	0	0	0	0	0	0	0	0	0	0	0	0	0	0	0	0	0	0	0	0	0	0	0	0	1	0	0	0	0	0	0	0	0	0	0	0	0	0	0	0	0	0	0	0	0
37	0	46	4	0	31	0	1	17	22	6	0	0	0	0	0	0	0	0	0	0	0	0	0	0	0	0	0	0	0	0	0	0	0	0	0	0	0	0	0	0	0	0	0	0	0	0	0	0	0	0	0	0	0	0	0
38	0	0	0	0	0	0	0	0	0	0	0	0	0	0	0	0	0	0	0	0	0	0	0	0	0	0	0	0	0	0	0	0	0	0	0	0	0	0	0	0	0	0	0	0	0	0	0	0	0	0	0	0	0	0	0
39	0	0	0	0	0	0	0	0	0	0	0	0	0	0	0	0	0	0	0	0	0	0	0	0	0	0	0	0	0	0	0	0	0	0	0	0	0	0	0	0	0	0	0	0	0	0	0	0	0	0	0	0	0	0	0
91	0	0	0	0	0	0	0	0	0	0	0	0	0	0	0	0	0	0	0	0	0	0	0	0	0	0	0	0	0	0	0	0	0	0	0	0	0	0	0	0	0	0	0	0	0	0	0	0	0	0	0	0	0	0	0

The Caviar Network (Complete/Part 3: Lower-Left Quarter)

	1	2	83	3	85	82	88	89	4	5	90	86	6	7	8	9	10	11	12	76	84	13	14	15	16	17	18	77	19	78	20	87	21	22	79	23	24	25	26	27	28	29	30	31	32	33	80	34	35	36	81	37	38	39	91
40	0	0	0	0	0	0	0	0	0	0	0	0	0	0	0	0	0	0	0	0	0	0	0	0	0	0	0	0	0	0	0	0	0	0	0	0	0	0	0	0	0	0	0	0	0	0	0	0	0	0	0	0	0	0	0
41	0	0	0	0	0	0	0	0	0	0	0	0	0	0	0	0	0	0	2	0	0	10	0	0	0	0	0	0	0	0	0	0	0	0	0	0	0	0	0	4	0	0	0	0	0	0	0	0	0	0	0	12	0	0	0
42	0	0	0	0	0	0	0	0	0	0	0	0	0	0	0	0	0	0	2	0	0	0	0	0	0	0	0	0	0	0	0	0	0	0	0	0	0	0	0	0	0	0	0	0	0	0	0	0	0	0	0	0	0	0	0
92	0	0	0	0	0	0	0	0	0	0	0	0	0	0	0	0	0	0	0	0	0	0	0	0	0	0	0	0	0	0	0	0	0	0	0	0	0	0	0	0	0	0	0	0	0	0	0	0	0	0	0	0	0	0	0
93	0	0	0	0	0	0	0	0	0	0	0	0	0	0	0	0	0	0	0	0	0	0	0	0	0	0	0	0	0	0	0	0	0	0	0	0	0	0	0	0	0	0	0	0	0	0	0	0	0	0	0	1	0	0	0
94	0	0	0	0	0	0	0	0	0	0	0	0	0	0	0	0	0	0	0	0	0	0	0	0	0	0	0	0	0	0	0	0	0	0	0	0	0	0	0	0	0	0	0	0	0	0	0	0	0	0	0	0	0	0	0
43	0	0	0	0	0	0	0	0	0	0	0	0	0	0	0	0	0	0	0	0	0	0	0	0	0	0	0	0	0	0	0	0	0	0	0	0	0	0	0	0	0	0	0	0	0	0	0	0	0	0	0	1	0	0	0
95	0	0	0	0	0	0	0	0	0	0	0	0	0	0	0	0	0	0	0	0	0	0	0	0	0	0	0	0	0	0	0	0	0	0	0	0	0	0	0	0	0	0	0	0	0	0	0	0	0	0	0	0	0	0	0
44	0	0	0	0	0	0	0	0	0	0	0	0	0	0	0	0	0	0	0	0	0	0	0	0	0	0	0	0	0	0	0	0	0	0	0	0	0	0	0	0	0	0	0	0	0	0	0	0	0	0	0	2	0	0	0
45	0	0	0	0	0	0	0	0	0	0	0	0	0	0	0	0	0	0	0	0	0	0	0	0	0	0	0	0	0	0	0	0	0	0	0	0	0	0	0	0	0	0	0	0	0	0	0	0	0	0	0	1	0	0	0
96	1	0	0	0	0	7	0	0	0	0	0	0	0	0	0	0	0	0	0	0	0	0	0	0	0	0	0	0	0	0	0	0	0	0	0	0	0	0	0	0	0	0	0	0	0	0	0	0	0	0	0	0	0	0	0
46	0	0	0	0	0	8	0	0	0	0	0	0	0	0	0	0	0	0	0	0	0	0	0	0	0	0	0	0	0	0	0	0	0	0	0	0	0	0	0	0	0	0	0	0	0	0	0	0	0	0	0	0	0	0	0
97	0	0	0	0	0	0	0	0	0	0	0	0	0	0	0	0	0	0	0	0	0	0	0	0	0	0	0	0	0	0	0	0	0	0	0	0	0	0	0	0	0	0	0	0	0	0	0	0	0	0	0	0	0	0	0
47	2	0	0	0	0	0	0	0	0	0	0	0	0	0	0	0	0	0	0	0	0	0	0	0	0	0	0	0	0	0	0	0	0	0	0	0	0	0	0	0	0	0	0	0	0	0	0	0	0	0	0	0	0	0	0
98	1	0	0	0	0	0	0	0	0	0	0	0	0	0	0	0	0	0	0	0	0	0	0	0	0	0	0	0	0	0	0	0	0	0	0	0	0	0	0	0	0	0	0	0	0	0	0	0	0	0	0	0	0	0	0
48	0	0	0	0	0	0	0	0	0	0	0	0	0	0	0	0	0	0	0	0	0	0	0	0	0	0	0	0	0	0	0	0	0	0	0	0	0	0	0	0	0	0	0	0	0	0	0	0	0	0	0	0	0	0	0
49	1	0	2	1	0	0	0	0	0	0	0	1	0	0	0	0	0	0	0	0	0	0	0	0	0	0	0	0	0	0	0	0	0	0	0	0	0	0	0	0	0	0	0	0	0	0	0	0	0	0	0	0	0	0	0
50	1	0	0	0	0	0	0	0	0	0	0	0	0	0	0	0	0	0	0	0	0	0	0	0	0	0	0	0	0	0	0	0	0	0	0	0	0	0	0	0	0	0	0	0	0	0	0	0	0	0	0	0	0	0	0
51	0	0	0	1	0	0	0	0	0	0	0	0	0	0	0	1	0	0	0	0	0	0	0	0	0	0	0	0	0	0	0	0	0	0	0	0	0	0	0	0	0	0	0	0	0	0	0	0	0	0	0	0	0	0	0
99	0	0	0	1	0	0	0	0	0	0	0	0	0	0	0	0	0	0	0	0	0	0	0	0	0	0	0	0	0	0	0	0	0	0	0	0	0	0	0	0	0	0	0	0	0	0	0	0	0	0	0	0	0	0	0
52	0	0	4	0	0	0	0	0	0	0	0	0	0	0	0	0	0	0	0	0	0	0	0	0	0	0	0	0	0	0	0	0	0	0	0	0	0	0	0	0	0	0	0	0	0	0	0	0	0	0	0	0	0	0	0
53	1	0	0	0	0	0	0	0	0	0	0	0	0	0	0	0	0	0	0	0	0	0	0	0	0	0	0	0	0	0	0	0	0	0	0	0	0	0	0	0	0	0	0	0	0	0	0	0	0	0	0	0	0	0	0
54	2	0	0	0	0	0	0	0	0	0	0	0	0	0	0	0	0	0	0	0	0	0	0	0	0	0	0	0	0	0	0	0	0	0	0	0	0	0	0	0	0	0	0	0	0	0	0	0	0	0	0	0	0	0	0
55	0	0	0	0	0	0	0	0	0	0	0	0	0	0	0	0	0	0	0	0	0	0	0	0	0	0	0	0	0	0	0	0	0	0	0	0	0	0	0	0	0	0	0	0	0	0	0	0	0	0	0	0	0	0	0
56	0	0	0	0	0	0	0	0	0	0	0	0	0	0	0	0	0	0	0	0	0	0	0	0	0	0	0	0	0	0	0	0	0	0	0	0	0	0	0	0	0	0	0	0	0	0	0	0	0	0	0	0	0	0	0
100	1	0	0	0	0	0	0	0	0	0	0	0	0	0	0	0	0	0	0	0	0	0	0	0	0	0	0	0	0	0	0	0	0	0	0	0	0	0	0	0	0	0	0	0	0	0	0	0	0	0	0	0	0	0	0
57	0	0	0	0	0	0	0	0	0	0	0	0	0	0	0	0	0	0	0	0	0	0	0	0	0	0	0	0	0	0	0	0	0	0	0	0	0	0	0	0	0	0	0	0	0	0	0	0	0	0	0	0	0	0	0
101	0	0	0	0	0	0	0	0	0	0	0	0	0	0	0	0	0	0	0	0	0	0	0	0	0	0	0	0	0	0	0	0	0	0	11	0	0	0	0	0	0	0	0	0	0	0	0	0	0	0	0	0	0	0	0
58	0	0	0	0	0	0	0	0	0	0	0	0	0	0	0	0	0	0	0	0	0	0	26	0	0	0	0	0	0	0	0	0	0	0	0	0	6	0	0	0	0	0	0	0	0	0	0	0	0	0	0	0	0	0	0
59	0	0	0	0	0	0	0	0	0	0	0	0	0	0	0	0	0	0	0	0	0	0	0	0	0	0	0	0	0	0	0	0	0	0	4	0	0	0	0	0	0	0	0	0	0	0	0	0	0	0	0	0	0	0	0
102	0	0	0	0	0	0	0	0	0	0	0	0	0	0	0	0	0	0	0	0	0	0	0	0	0	0	0	0	0	0	0	0	0	0	2	0	0	0	0	0	0	0	0	0	0	0	0	0	0	0	0	0	0	0	0
103	0	0	0	0	0	0	0	0	0	0	0	0	0	0	0	0	0	0	0	0	0	0	0	0	0	0	0	0	0	0	0	0	0	0	0	0	0	0	0	0	0	0	0	0	0	0	0	0	0	0	0	0	0	0	0
104	0	0	0	0	0	0	0	0	0	0	0	0	0	0	0	0	0	0	0	0	0	0	0	0	0	0	0	0	0	0	0	0	0	0	0	0	0	0	0	0	0	0	0	0	0	0	0	0	0	0	0	0	0	0	0
60	0	0	0	0	0	0	0	0	0	0	0	0	0	0	0	0	0	0	0	0	0	0	0	0	0	0	0	0	0	0	0	0	0	0	0	0	0	0	0	0	0	0	0	0	0	0	0	0	0	0	9	0	0	0	0
105	0	0	0	0	0	0	0	0	0	0	0	0	0	0	0	0	0	0	0	0	0	0	0	0	0	0	0	0	0	0	0	0	0	0	0	0	0	0	0	0	0	0	0	0	0	0	0	0	0	0	0	0	0	0	0
61	0	0	0	0	0	0	0	0	0	0	0	0	0	0	0	0	0	0	0	1	0	0	0	0	0	0	0	0	0	0	0	0	0	0	0	0	0	0	0	0	0	0	0	0	0	0	0	0	0	0	0	0	0	0	0
62	0	0	0	0	0	0	0	0	0	0	0	0	0	0	0	0	0	0	0	0	0	0	0	0	0	0	0	0	0	0	0	0	0	0	0	0	0	0	0	0	0	0	0	0	0	0	0	0	0	0	0	0	0	0	0
106	0	0	0	0	0	0	0	0	0	0	0	0	0	0	0	0	0	0	0	0	0	0	0	0	0	0	0	0	0	0	0	0	0	0	0	0	0	0	0	0	0	0	0	0	0	0	0	0	0	0	0	0	0	0	0
107	0	0	0	0	0	0	0	0	0	0	0	0	0	0	0	0	0	0	0	0	0	0	0	0	0	0	0	0	0	0	0	0	0	0	0	0	0	0	0	0	0	0	0	0	0	0	0	0	0	0	0	0	0	0	0
108	4	0	0	0	0	0	0	0	0	0	0	0	1	0	0	0	0	0	0	0	0	0	0	0	0	0	0	0	0	0	0	0	0	0	0	0	0	0	0	0	0	0	0	0	0	0	0	0	0	0	0	0	0	0	0
63	0	0	0	0	0	0	0	0	0	0	0	0	0	0	0	0	0	0	0	0	0	0	0	0	0	0	0	0	0	0	0	0	0	0	0	0	0	0	0	0	0	0	0	0	0	0	0	0	0	0	0	0	0	0	0
109	0	0	0	0	0	0	0	0	0	0	0	0	0	0	0	0	0	0	0	0	0	0	0	0	0	0	0	0	0	0	0	0	0	0	0	0	0	0	0	0	0	0	0	0	0	0	0	0	0	0	0	0	0	0	0
110	0	0	0	0	0	0	0	0	0	0	0	0	0	0	0	0	0	0	0	0	0	0	0	0	0	0	0	0	0	0	0	0	0	0	0	0	0	0	0	0	0	0	0	0	0	0	0	0	0	0	0	0	0	0	0
64	0	0	0	0	0	0	0	0	0	0	0	0	0	0	0	0	0	0	0	0	0	0	0	0	0	0	0	0	0	0	0	0	0	0	0	0	0	0	0	0	0	0	0	0	0	0	0	0	0	0	0	0	0	0	0
65	0	0	0	0	0	0	0	0	0	0	0	0	0	0	0	0	0	0	4	0	0	0	0	0	0	0	0	0	0	0	0	0	0	0	0	0	0	0	0	0	0	0	0	0	0	0	0	0	0	0	0	0	0	0	0
66	0	0	0	0	0	0	0	0	0	0	0	0	0	0	0	0	0	0	2	0	0	0	0	0	0	0	0	0	0	0	0	0	0	0	0	0	0	0	0	0	0	0	0	0	0	0	0	0	0	0	0	0	0	0	0
67	0	0	0	0	0	0	0	0	0	0	0	0	0	0	0	0	0	0	0	0	0	0	0	0	0	0	0	0	0	0	0	0	0	0	0	0	0	0	0	0	0	0	0	0	0	0	0	0	0	0	0	0	0	0	0
68	11	0	0	0	0	0	0	0	0	0	0	0	0	0	0	0	0	0	0	0	0	0	0	0	0	0	0	0	0	0	0	0	0	0	0	0	0	0	0	0	0	0	0	0	0	0	0	0	0	0	0	0	0	0	0
69	11	0	0	0	0	0	0	0	0	0	0	0	0	0	0	0	0	0	0	0	0	0	0	0	0	0	0	0	0	0	0	0	0	0	0	0	0	0	0	0	0	0	0	0	0	0	0	0	0	0	0	0	0	0	0
70	0	0	0	0	0	0	0	0	0	0	0	0	0	0	0	0	0	0	0	0	0	0	0	0	0	0	0	0	0	0	0	0	0	0	0	0	0	0	0	0	0	0	0	0	0	0	0	0	0	0	0	0	0	0	0
71	0	0	0	0	0	0	0	0	0	0	0	0	0	0	0	0	0	0	0	3	0	0	0	0	0	0	0	0	0	0	0	0	0	0	0	0	0	0	0	0	0	0	0	0	0	0	0	0	0	0	0	0	0	0	0
72	0	0	0	0	0	0	0	0	0	0	0	0	0	0	0	0	0	0	0	0	0	0	0	0	0	0	0	0	0	0	0	0	0	0	0	0	0	0	0	0	0	0	0	0	0	0	0	0	0	0	0	0	0	0	0
73	0	0	0	0	0	0	0	0	0	0	0	0	0	0	0	0	0	0	0	0	0	0	0	0	0	0	0	0	0	0	0	0	0	0	0	0	0	0	0	0	0	0	0	0	0	0	0	0	0	0	0	0	0	0	0
74	0	0	0	0	0	0	0	0	0	0	0	0	0	0	0	0	0	0	0	0	0	0	0	0	0	0	0	0	0	0	0	0	0	0	0	0	0	0	0	0	0	0	0	0	0	0	0	0	0	0	0	0	0	0	0
75	0	0	0	0	0	0	0	0	0	0	0	0	0	0	0	0	0	0	1	0	0	0	0	0	0	0	0	0	0	0	0	0	0	0	0	0	0	0	0	0	0	0	0	0	0	0	0	0	0	0	0	0	0	0	0

The Caviar Network (Complete/Part 4: Lower-Left Quarter)

	40	41	42	92	93	94	43	95	44	45	96	46	97	47	98	48	49	50	51	99	52	53	54	55	56	100	57	101	58	59	102	103	104	60	105	61	62	106	107	108	63	109	110	64	65	66	67	68	69	70	71	72	73	74	75
40	0	0	0	0	0	0	0	0	0	0	0	0	0	0	0	0	0	0	0	0	0	0	0	0	0	0	0	0	0	0	0	0	0	0	0	0	0	0	0	0	0	0	0	0	0	0	0	0	0	0	0	0	0	0	0
41	0	0	1	11	3	10	0	0	0	0	0	0	0	0	0	0	0	0	0	0	0	0	0	0	0	0	0	0	0	0	0	0	0	0	0	0	0	0	0	0	0	0	0	0	0	0	0	0	0	0	0	0	0	0	0
42	0	0	0	0	0	0	0	0	0	0	0	0	0	0	0	0	0	0	0	0	0	0	0	0	0	0	0	0	0	0	0	0	0	0	0	0	0	0	0	0	0	0	0	0	0	0	0	0	0	0	0	0	0	0	0
92	0	0	0	0	0	0	0	0	0	0	0	0	0	0	0	0	0	0	0	0	0	0	0	0	0	0	0	0	0	0	0	0	0	0	0	0	0	0	0	0	0	0	0	0	0	0	0	0	0	0	0	0	0	0	0
93	0	1	0	0	0	0	0	0	0	0	0	0	0	0	0	0	0	0	0	0	0	0	0	0	0	0	0	0	0	0	0	0	0	0	0	0	0	0	0	0	0	0	0	0	0	0	0	0	0	0	0	0	0	0	0
94	0	0	0	0	0	0	0	0	0	0	0	0	0	0	0	0	0	0	0	0	0	0	0	0	0	0	0	0	0	0	0	0	0	0	0	0	0	0	0	0	0	0	0	0	0	0	0	0	0	0	0	0	0	0	0
43	0	1	0	0	0	0	0	0	0	0	0	0	0	0	0	0	0	0	0	0	0	0	0	0	0	0	0	0	0	0	0	0	0	0	0	0	0	0	0	0	0	0	0	0	0	0	0	0	0	0	0	0	0	0	0
95	0	0	0	0	0	0	0	0	0	0	0	0	0	0	0	0	0	0	0	0	0	0	0	0	0	0	0	0	0	0	0	0	0	0	0	0	0	0	0	0	0	0	0	0	0	0	0	0	0	0	0	0	0	0	0
44	0	0	0	0	0	0	0	0	0	0	0	0	0	0	0	0	0	0	0	0	0	0	0	0	0	0	0	0	0	0	0	0	0	0	0	0	0	0	0	0	0	0	0	0	0	0	0	0	0	0	0	0	0	0	0
45	0	0	0	0	0	0	0	0	0	0	0	0	0	0	0	0	0	0	0	0	0	0	0	0	0	0	0	0	0	0	0	0	0	0	0	0	0	0	0	0	0	0	0	0	0	0	0	0	0	0	0	0	0	0	0
96	0	0	0	0	0	0	0	0	0	0	0	8	0	0	0	0	0	0	0	0	0	0	0	0	0	0	0	0	0	0	0	0	0	0	0	0	0	0	0	0	0	0	0	0	0	0	0	0	0	0	0	0	0	0	0
46	0	0	0	0	0	0	0	0	0	0	5	0	0	0	0	0	0	0	0	0	0	0	0	0	0	0	0	0	0	0	0	0	0	0	0	0	0	0	0	0	0	0	0	0	0	0	0	0	0	0	0	0	0	0	0
97	0	0	0	0	0	0	0	0	0	0	0	0	0	0	0	0	0	0	0	0	0	0	0	0	0	0	0	0	0	0	0	0	0	0	0	0	0	0	0	0	0	0	0	0	0	0	0	0	0	0	0	0	0	0	0
47	0	0	0	0	0	0	0	0	0	0	0	0	0	0	0	0	0	0	0	0	0	0	0	0	0	0	0	0	0	0	0	0	0	0	0	0	0	0	0	0	0	0	0	0	0	0	0	0	0	0	0	0	0	0	0
98	0	0	0	0	0	0	0	0	0	0	0	0	0	0	0	0	0	0	0	0	0	0	0	0	0	0	0	0	0	0	0	0	0	0	0	0	0	0	0	0	0	0	0	0	0	0	0	0	0	0	0	0	0	0	0
48	0	0	0	0	0	0	0	0	0	0	0	0	0	0	0	0	0	0	0	0	0	0	0	0	0	0	0	0	0	0	0	0	0	0	0	0	0	0	0	0	0	0	0	0	0	0	0	0	0	0	0	0	0	0	0
49	0	0	0	0	0	0	0	0	0	0	0	0	0	0	0	0	0	0	0	0	0	0	0	0	0	0	0	0	0	0	0	0	0	0	0	0	0	0	0	0	0	0	0	0	0	0	0	0	0	0	0	0	0	0	0
50	0	0	0	0	0	0	0	0	0	0	0	0	0	0	0	0	0	0	0	0	0	0	0	0	0	0	0	0	0	0	0	0	0	0	0	0	0	0	0	0	0	0	0	0	0	0	0	0	0	0	0	0	0	0	0
51	0	0	0	0	0	0	0	0	0	0	0	0	0	0	0	0	0	0	0	0	0	0	0	0	0	0	0	0	0	0	0	0	0	0	0	0	0	0	0	0	0	0	0	0	0	0	0	0	0	0	0	0	0	0	0
99	0	0	0	0	0	0	0	0	0	0	0	0	0	0	0	0	0	0	0	0	0	0	0	0	0	0	0	0	0	0	0	0	0	0	0	0	0	0	0	0	0	0	0	0	0	0	0	0	0	0	0	0	0	0	0
52	0	0	0	0	0	0	0	0	0	0	0	0	0	0	0	0	0	0	0	0	0	0	0	0	0	0	0	0	0	0	0	0	0	0	0	0	0	0	0	0	0	0	0	0	0	0	0	0	0	0	0	0	0	0	0
53	0	0	0	0	0	0	0	0	0	0	0	0	0	0	0	0	0	0	0	0	0	0	0	0	0	0	0	0	0	0	0	0	0	0	0	0	0	0	0	0	0	0	0	0	0	0	0	0	0	0	0	0	0	0	0
54	0	0	0	0	0	0	0	0	0	0	0	0	0	0	0	0	0	0	0	0	0	0	0	0	0	0	0	0	0	0	0	0	0	0	0	0	0	0	0	0	0	0	0	0	0	0	0	0	0	0	0	0	0	0	0
55	0	0	0	0	0	0	0	0	0	0	0	0	0	0	0	0	0	0	0	0	0	0	0	0	0	0	0	0	0	0	0	0	0	0	0	0	0	0	0	0	0	0	0	0	0	0	0	0	0	0	0	0	0	0	0
56	0	0	0	0	0	0	0	0	0	0	0	0	0	0	0	0	0	0	0	0	0	0	0	0	0	0	0	0	0	0	0	0	0	0	0	0	0	0	0	0	0	0	0	0	0	0	0	0	0	0	0	0	0	0	0
100	0	0	0	0	0	0	0	0	0	0	0	0	0	0	0	0	0	0	0	0	0	0	0	0	0	0	0	0	0	0	0	0	0	0	0	0	0	0	0	0	0	0	0	0	0	0	0	0	0	0	0	0	0	0	0
57	0	0	0	0	0	0	0	0	0	0	0	0	0	0	0	0	0	0	0	0	0	0	0	0	0	0	0	0	0	0	0	0	0	0	0	0	0	0	0	0	0	0	0	0	0	0	0	0	0	0	0	0	0	0	0
101	0	0	0	0	0	0	0	0	0	0	0	0	0	0	0	0	0	0	0	0	0	0	0	0	0	0	0	0	0	0	3	0	0	0	0	0	0	0	0	0	0	0	0	0	0	0	0	0	0	0	0	0	0	0	0
58	0	0	0	0	0	0	0	0	0	0	0	0	0	0	0	0	0	0	0	0	0	0	0	0	0	0	0	0	0	0	0	0	0	0	0	0	0	0	0	0	0	0	0	0	0	0	0	0	0	0	0	0	0	0	0
59	0	0	0	0	0	0	0	0	0	0	0	0	0	0	0	0	0	0	0	0	0	0	0	0	0	0	0	0	0	0	0	0	0	0	0	0	0	0	0	0	0	0	0	0	0	0	0	0	0	0	0	0	0	0	0
102	0	0	0	0	0	0	0	0	0	0	0	0	0	0	0	0	0	0	0	0	0	0	0	0	0	0	0	0	0	1	0	0	0	0	0	0	0	0	0	0	0	0	0	0	0	0	0	0	0	0	0	0	0	0	0
103	0	0	0	0	0	0	0	0	0	0	0	0	0	0	0	0	0	0	0	0	0	0	0	0	0	0	0	0	0	0	0	0	0	0	0	0	0	0	0	0	0	0	0	0	0	0	0	0	0	0	0	0	0	0	0
104	0	0	0	0	0	0	0	0	0	0	0	0	0	0	0	0	0	0	0	0	0	0	0	0	0	0	0	0	0	0	0	1	0	0	0	0	0	0	0	0	0	0	0	0	0	0	0	0	0	0	0	0	0	0	0
60	0	0	0	0	0	0	0	0	0	0	0	0	0	0	0	0	0	0	0	0	0	0	0	0	0	0	0	0	0	0	0	0	0	0	0	0	0	0	0	0	0	0	0	0	0	0	0	0	0	0	0	0	0	0	0
105	0	0	0	0	0	0	0	0	0	0	0	0	0	0	0	0	0	0	0	0	0	0	0	0	0	0	0	0	0	0	0	0	0	0	0	0	0	0	0	0	0	0	0	0	0	0	0	0	0	0	0	0	0	0	0
61	0	0	0	0	0	0	0	0	0	0	0	0	0	0	0	0	0	0	0	0	0	0	0	0	0	0	0	0	0	0	0	0	0	0	0	0	0	0	0	0	0	0	0	0	0	0	0	0	0	0	0	0	0	0	0
62	0	0	0	0	0	0	0	0	0	0	0	0	0	0	0	0	0	0	0	0	0	0	0	0	0	0	0	0	0	0	0	0	0	0	0	0	0	0	0	0	0	0	0	0	0	0	0	0	0	0	0	0	0	0	0
106	0	0	0	0	0	0	0	0	0	0	0	0	0	0	0	0	0	0	0	0	0	0	0	0	0	0	0	0	0	0	0	0	0	0	0	0	0	0	0	0	0	0	0	0	0	0	0	0	0	0	0	0	0	0	0
107	0	0	0	0	0	0	0	0	0	0	0	0	0	0	0	0	0	0	0	0	0	0	0	0	0	0	0	0	0	0	0	0	0	0	0	0	0	0	0	0	3	0	0	0	0	0	0	0	0	0	0	0	0	0	0
108	0	0	0	0	0	0	0	0	0	0	0	0	0	0	0	0	0	0	0	0	0	0	0	0	0	0	0	0	0	0	0	0	0	0	0	0	0	0	0	0	0	0	0	0	0	0	0	0	0	0	0	0	0	0	0
63	0	0	0	0	0	0	0	0	0	0	0	0	0	0	0	0	0	0	0	0	0	0	0	0	0	0	0	0	0	0	0	0	0	0	0	0	0	0	0	0	0	0	0	0	0	0	0	0	0	0	0	0	0	0	0
109	0	0	0	0	0	0	0	0	0	0	0	0	0	0	0	0	0	0	0	0	0	0	0	0	0	0	0	0	0	0	0	0	0	0	0	0	0	0	0	0	0	0	0	0	0	0	0	0	0	0	0	0	0	0	0
110	0	0	0	0	0	0	0	0	0	0	0	0	0	0	0	0	0	0	0	0	0	0	0	0	0	0	0	0	0	0	0	0	0	0	0	0	0	0	0	0	0	0	0	0	0	0	0	0	0	0	0	0	0	0	0
64	0	0	0	0	0	0	0	0	0	0	0	0	0	0	0	0	0	0	0	0	0	0	0	0	0	0	0	0	0	0	0	0	0	0	0	0	0	0	0	0	0	0	0	0	0	0	0	0	0	0	0	0	0	0	0
65	0	0	0	0	0	0	0	0	0	0	0	0	0	0	0	0	0	0	0	0	0	0	0	0	0	0	0	0	0	0	0	0	0	0	0	0	0	0	0	0	0	0	0	0	0	0	0	0	0	0	0	0	0	0	0
66	0	0	0	0	0	0	0	0	0	0	0	0	0	0	0	0	0	0	0	0	0	0	0	0	0	0	0	0	0	0	0	0	0	0	0	0	0	0	0	0	0	0	0	0	0	0	0	0	0	0	0	0	0	0	0
67	0	0	0	0	0	0	0	0	0	0	0	0	0	0	0	0	0	0	0	0	0	0	0	0	0	0	0	0	0	0	0	0	0	0	0	0	0	0	0	0	0	0	0	0	0	0	0	0	0	0	0	0	0	0	0
68	0	0	0	0	0	0	0	0	0	0	0	0	0	0	0	0	0	0	0	0	0	0	0	0	0	0	0	0	0	0	0	0	0	0	0	0	0	0	0	0	0	0	0	0	0	0	0	0	0	0	0	0	0	0	0
69	0	0	0	0	0	0	0	0	0	0	0	0	0	0	0	0	0	0	0	0	0	0	0	0	0	0	0	0	0	0	0	0	0	0	0	0	0	0	0	0	0	0	0	0	0	0	0	0	0	0	0	0	0	0	0
70	0	0	0	0	0	0	0	0	0	0	0	0	0	0	0	0	0	0	0	0	0	0	0	0	0	0	0	0	0	0	0	0	0	0	0	0	0	0	0	0	0	0	0	0	0	0	0	0	0	0	0	0	0	0	0
71	0	0	0	0	0	0	0	0	0	0	0	0	0	0	0	0	0	0	0	0	0	0	0	0	0	0	0	0	0	0	0	0	0	0	0	0	0	0	0	0	0	0	0	0	0	0	0	0	0	0	0	0	0	0	0
72	0	0	0	0	0	0	0	0	0	0	0	0	0	0	0	0	0	0	0	0	0	0	0	0	0	0	0	0	0	0	0	0	0	0	0	0	0	0	0	0	0	0	0	0	0	0	0	0	0	0	0	0	0	0	0
73	0	0	0	0	0	0	0	0	0	0	0	0	0	0	0	0	0	0	0	0	0	0	0	0	0	0	0	0	0	0	0	0	0	0	0	0	0	0	0	0	0	0	0	0	0	0	0	0	0	0	0	0	0	0	0
74	0	0	0	0	0	0	0	0	0	0	0	0	0	0	0	0	0	0	0	0	0	0	0	0	0	0	0	0	0	0	0	0	0	0	0	0	0	0	0	0	0	0	0	0	0	0	0	0	0	0	0	0	0	0	0
75	0	0	0	0	0	0	0	0	0	0	0	0	0	0	0	0	0	0	0	0	0	0	0	0	0	0	0	0	0	0	0	0	0	0	0	0	0	0	0	0	0	0	0	0	0	0	0	0	0	0	0	0	0	0	0

The Caviar Network (Phase 1)

	1	4	89	83	3	5	88	85	90	2	7	54	6	64	8
1	0	1	4	0	4	2	2	9	1	2	0	2	0	1	1
4	0	0	0	0	0	0	0	0	0	0	0	0	0	0	0
89	1	0	0	0	0	0	0	0	0	0	3	0	0	0	0
83	1	0	0	0	0	0	0	0	0	0	0	0	5	0	0
3	2	0	0	0	0	0	1	0	0	0	0	0	0	0	0
5	0	0	0	0	0	0	0	0	0	0	0	0	0	0	0
88	1	0	0	0	1	0	0	3	0	0	0	0	1	0	0
85	1	0	0	0	0	0	2	0	0	0	0	0	5	0	0
90	0	0	0	0	0	0	0	0	0	0	0	0	0	0	0
2	0	0	0	0	0	0	0	0	0	0	0	0	0	0	0
7	0	0	2	0	0	0	0	0	0	0	0	0	0	0	0
54	0	0	0	0	0	0	0	0	0	0	0	0	0	0	0
6	0	0	0	0	0	0	1	0	0	0	0	0	0	0	0
64	0	0	0	0	0	0	0	0	0	0	0	0	0	0	0
8	0	0	0	0	0	0	0	0	0	0	0	0	0	0	0

The Caviar Network (Phase 2)

	1	89	83	3	5	88	85	90	86	2	7	6	64	8	55	10	56	97	47	98	76	9	11	12
1	0	6	8	31	5	1	0	3	0	7	0	0	3	7	4	3	1	2	1	1	4	5	0	0
89	0	0	0	0	0	0	0	0	0	0	1	0	0	0	0	0	0	0	0	0	0	0	1	0
83	0	0	0	0	0	0	0	0	0	0	0	0	0	0	0	0	0	0	0	0	0	0	0	0
3	0	0	0	0	0	0	0	0	0	0	0	0	0	1	0	1	0	0	0	0	0	0	0	0
5	0	0	0	0	0	0	0	0	0	0	0	0	0	0	0	0	0	0	0	0	0	0	0	0
88	1	0	0	0	0	0	0	0	0	0	0	0	0	0	0	0	0	0	0	0	0	0	1	0
85	0	0	0	0	0	1	0	0	0	0	0	0	0	0	0	0	0	0	0	0	1	0	0	0
90	0	0	0	0	0	0	0	0	0	0	0	0	0	0	0	0	0	0	0	0	0	0	0	0
86	1	0	0	0	0	0	0	0	0	0	0	0	0	0	0	0	0	0	0	0	0	0	0	0
2	5	0	0	0	0	0	0	0	0	0	0	0	0	0	0	0	0	0	0	0	0	0	0	0
7	0	1	0	0	0	0	0	0	0	0	0	0	0	0	0	0	0	0	0	0	0	0	0	0
6	0	0	1	0	0	0	0	0	0	0	0	0	0	0	0	0	0	0	0	0	0	0	0	0
64	0	0	0	0	0	0	0	0	0	0	0	0	0	0	0	0	0	0	0	0	0	0	0	0
8	5	0	0	0	0	0	0	0	0	0	0	0	2	0	0	0	0	0	0	0	0	0	0	0
55	1	0	0	0	0	0	0	0	0	0	0	0	0	0	0	0	0	0	0	0	0	0	0	0
10	0	0	0	0	0	0	0	0	0	0	0	0	0	0	0	0	0	0	0	0	0	0	0	0
56	0	0	0	0	0	0	0	0	0	0	0	0	0	0	0	0	0	0	0	0	0	0	0	0
97	0	0	0	0	0	0	0	0	0	0	0	0	0	0	0	0	0	0	0	0	0	0	0	0
47	1	0	0	0	0	0	0	0	0	0	0	0	0	0	0	0	0	0	0	0	0	0	0	0
98	1	0	0	0	0	0	0	0	0	0	0	0	0	0	0	0	0	0	0	0	0	0	0	0
76	0	0	0	0	0	0	0	0	0	0	0	0	0	0	0	0	0	0	0	0	0	0	0	0
9	0	0	0	0	0	0	0	0	0	0	0	0	0	0	0	0	0	0	0	0	0	0	0	0
11	0	0	0	0	0	0	0	0	0	0	0	0	0	0	0	0	0	0	0	0	0	0	0	0
12	2	0	0	0	0	0	0	0	0	0	0	0	0	0	0	0	0	0	0	0	0	0	0	0

The Caviar Network (Phase 3)

	1	4	89	83	3	5	48	88	85	90	86	2	7	6	8	55	10	56	76	9	34	35	11	32	84	49	107	50	99	13	51	12	52
1	0	0	7	38	29	7	2	0	11	2	29	8	1	4	3	3	3	2	5	17	3	1	0	1	4	3	0	3	0	2	0	2	1
4	0	0	0	0	0	1	0	0	0	0	0	0	0	0	0	0	0	0	0	0	0	0	0	0	0	0	0	0	0	0	0	0	0
89	0	0	0	0	0	0	0	0	0	0	0	0	1	0	0	0	0	0	0	0	0	0	0	0	0	0	0	0	0	0	0	0	0
83	4	0	0	0	0	0	0	0	0	0	9	0	0	2	0	0	0	0	0	0	0	0	1	0	3	0	3	0	0	0	0	0	2
3	1	0	0	0	0	0	0	0	4	0	3	0	0	0	0	0	0	0	0	5	0	0	0	0	0	0	0	0	1	0	0	2	0
5	0	0	0	0	0	0	0	0	0	0	0	0	0	0	0	0	0	0	0	0	0	0	0	0	0	0	0	0	0	0	0	0	0
48	0	0	0	0	0	0	0	0	0	0	0	0	0	0	0	0	0	0	0	0	0	0	0	0	0	0	0	0	0	0	0	0	0
88	0	0	0	0	0	0	0	0	3	0	0	0	0	1	0	0	0	0	0	0	0	0	0	0	0	0	0	0	0	0	0	0	0
85	0	0	0	0	0	0	0	0	0	0	0	0	0	0	0	0	0	0	0	0	0	0	0	0	0	0	0	0	0	0	0	0	0
90	0	0	0	0	0	0	0	0	0	0	0	0	0	0	0	0	0	0	0	0	0	0	0	0	0	0	0	0	0	0	0	0	0
86	3	0	0	2	1	0	0	0	0	0	0	0	0	0	0	0	0	0	0	0	0	0	0	0	0	0	1	0	0	0	0	0	0
2	0	0	0	0	0	0	0	0	0	0	0	0	0	0	0	0	0	0	0	0	1	0	0	1	0	0	0	0	0	0	0	0	0
7	0	0	0	0	0	0	0	0	0	0	0	0	0	0	0	0	0	0	0	0	0	0	0	0	0	0	0	0	0	0	0	0	0
6	3	0	0	0	1	0	0	0	0	0	0	0	0	0	0	0	0	0	0	0	0	0	0	0	0	0	0	0	0	0	0	0	0
8	0	0	0	0	0	0	0	0	0	0	0	0	0	0	0	0	0	0	0	0	0	0	0	0	0	0	0	0	0	0	0	0	0
55	0	0	0	0	0	0	0	0	0	0	0	0	0	0	0	0	0	0	0	0	0	0	0	0	0	0	0	0	0	0	0	0	0
10	0	0	0	0	0	0	0	0	0	0	0	0	0	0	0	0	0	0	0	0	0	0	0	0	0	0	0	0	0	0	0	0	0
56	0	0	0	0	0	0	0	0	0	0	0	0	0	0	0	0	0	0	0	0	0	0	0	0	0	0	0	0	0	0	0	0	0
76	1	0	0	0	0	0	0	0	2	0	0	0	0	0	0	0	0	0	0	0	0	0	0	0	0	0	0	0	0	0	0	0	0
9	1	0	0	0	0	0	0	0	0	0	0	0	0	0	1	0	0	0	0	0	0	0	0	0	0	0	0	0	0	0	0	0	0
34	0	0	0	0	0	0	0	0	0	0	0	0	0	0	0	0	0	0	0	0	0	0	0	0	0	0	0	0	0	0	0	0	0
35	0	0	0	0	0	0	0	0	0	0	0	0	0	0	0	0	0	0	0	0	0	0	0	0	0	0	0	0	0	0	0	0	0
11	6	0	0	0	1	0	0	0	0	0	0	0	0	0	0	0	0	0	0	0	0	0	0	0	0	0	0	0	0	0	0	0	0
32	0	0	0	0	0	0	0	0	0	0	0	0	0	0	0	0	0	0	0	0	0	0	0	0	0	0	0	0	0	0	0	0	0
84	4	0	0	0	0	0	0	0	0	0	0	0	0	0	0	0	0	0	0	0	0	0	0	0	0	0	0	0	0	0	0	0	2
49	2	0	0	3	1	0	0	0	0	1	0	0	0	0	0	0	0	0	0	0	0	0	0	0	2	0	0	0	0	0	0	0	0
107	0	0	0	0	0	0	0	0	0	0	0	0	0	0	0	0	0	0	0	0	0	0	0	0	0	0	0	0	0	0	0	0	0
50	1	0	0	0	0	0	0	0	0	0	0	0	0	0	0	0	0	0	0	0	0	0	0	0	0	0	0	0	0	0	0	0	0
99	0	0	0	0	1	0	0	0	0	0	0	0	0	0	0	0	0	0	0	0	0	0	0	0	0	0	0	0	0	0	0	0	0
13	0	0	0	0	0	0	0	0	0	0	0	0	0	0	0	0	0	0	0	2	0	0	0	0	0	0	0	0	0	0	0	0	0
51	0	0	0	0	0	0	0	0	0	0	0	0	0	0	0	0	0	0	0	1	0	0	0	0	0	0	0	0	0	0	0	0	0
12	0	0	0	0	0	0	0	0	0	0	0	0	0	0	0	0	0	0	0	0	0	0	0	0	0	0	0	0	0	0	0	0	0
52	0	0	0	0	0	0	0	0	0	0	0	0	0	0	0	0	0	0	0	0	0	0	0	0	0	0	0	0	0	0	0	0	0

The Caviar Network (Phase 4)

	1	4	89	83	3	5	88	85	90	86	2	7	6	8	47	76	9	35	11	53	84	49	107	13	51	63	109	31	12	14	52	106	15
1	0	1	12	17	71	16	7	45	1	3	16	0	11	10	1	9	15	0	3	0	0	1	0	7	0	0	0	0	0	0	2	1	7
4	0	0	0	0	0	0	0	0	0	0	0	0	0	0	0	0	0	0	0	0	0	0	0	0	0	0	0	0	0	0	0	0	0
89	1	0	0	0	0	0	0	0	0	0	0	2	0	0	0	0	0	0	0	0	0	1	0	0	0	0	2	1	0	1	0	0	0
83	0	0	0	0	0	0	0	1	0	8	0	0	0	0	0	0	0	0	0	0	1	0	1	0	0	0	0	0	0	0	2	1	0
3	9	0	0	0	0	0	0	3	0	0	0	0	0	1	0	0	6	0	0	0	0	0	0	0	0	0	0	7	2	0	0	0	0
5	3	0	0	0	0	0	0	0	0	0	0	0	0	0	0	0	0	0	0	0	0	0	0	0	0	0	0	0	0	0	0	0	0
88	0	0	0	0	0	0	0	2	0	0	0	0	0	0	0	0	0	0	0	0	0	0	0	0	0	0	0	0	0	0	0	0	0
85	7	0	0	0	2	0	6	0	0	0	0	0	0	0	0	3	0	0	0	0	0	0	0	0	0	0	0	0	0	0	0	0	0
90	1	0	0	0	1	0	0	0	0	0	0	0	0	0	0	0	0	0	0	0	0	0	0	0	0	0	0	0	0	0	0	0	0
86	3	0	0	9	0	0	0	0	0	0	0	0	0	0	0	0	0	0	0	0	0	0	2	0	0	0	0	0	0	0	0	0	0
2	3	0	0	0	0	0	0	0	0	0	0	0	0	0	0	0	0	0	0	0	0	0	0	0	0	0	0	0	0	0	0	0	0
7	0	0	3	0	0	0	0	0	0	0	0	0	0	0	0	0	0	0	0	0	0	0	0	0	0	0	0	0	0	0	0	0	0
6	15	0	0	0	0	0	0	0	0	0	0	0	0	0	0	0	0	0	0	0	0	0	0	0	0	0	0	0	0	0	0	0	0
8	0	0	0	0	0	0	0	0	0	0	0	0	0	0	0	0	4	1	0	0	0	0	0	0	0	0	0	0	0	0	0	0	0
47	0	0	0	0	0	0	0	0	0	0	0	0	0	0	0	0	0	0	0	0	0	0	0	0	0	0	0	0	0	0	0	0	0
76	0	0	0	0	0	0	0	6	0	0	0	0	0	0	0	0	0	0	0	0	0	0	0	0	0	0	0	0	0	0	0	0	0
9	0	0	0	0	0	0	0	0	0	0	0	0	0	0	0	0	0	0	0	0	0	0	0	0	0	0	0	0	0	0	0	0	0
35	0	0	0	0	0	0	0	0	0	0	0	0	0	0	0	0	0	0	0	0	0	0	0	0	0	0	0	0	0	0	0	0	0
11	29	0	0	0	0	0	0	0	0	0	0	0	0	0	0	0	0	0	0	0	0	0	0	0	0	0	0	0	0	0	0	0	0
53	1	0	0	0	0	0	0	0	0	0	0	0	0	0	0	0	0	0	0	0	0	0	0	0	0	0	0	0	0	0	0	0	0
84	1	0	0	1	0	0	0	0	0	0	0	0	0	0	0	0	0	0	0	0	0	0	0	0	0	0	0	0	0	0	0	0	0
49	0	0	0	0	0	0	0	0	0	0	0	0	0	0	0	0	0	0	0	0	0	0	0	0	0	0	0	0	0	0	0	0	0
107	0	0	0	0	0	0	0	0	0	0	0	0	0	0	0	0	0	0	0	0	0	0	0	0	0	40	0	0	0	0	0	0	0
13	0	0	0	0	0	0	0	0	0	0	0	0	0	0	0	0	0	0	0	0	0	0	0	0	1	0	0	0	0	0	0	0	0
51	0	0	0	0	0	0	0	0	0	0	0	0	0	0	0	0	0	0	0	0	0	0	0	0	0	0	0	0	0	0	0	0	0
63	0	0	0	0	0	0	0	0	0	0	0	0	0	0	0	0	0	0	0	0	0	0	0	0	0	0	0	0	0	0	0	0	0
109	0	0	0	0	0	0	0	0	0	0	0	0	0	0	0	0	0	0	0	0	0	0	0	0	0	0	0	0	0	0	0	0	0
31	0	0	0	0	0	0	0	0	0	0	0	0	0	0	0	0	1	0	0	0	0	0	0	0	0	0	0	0	0	0	0	0	0
12	0	0	0	0	1	0	0	0	0	0	0	0	0	0	0	0	0	0	0	0	0	0	0	0	0	0	0	0	0	0	0	0	0
14	0	0	0	0	0	0	0	0	0	0	0	0	0	0	0	0	0	0	0	0	0	0	0	0	0	0	0	0	0	0	0	0	0
52	0	0	0	4	0	0	0	0	0	0	0	0	0	0	0	0	0	0	0	0	0	0	0	0	0	0	0	0	0	0	0	0	0
106	0	0	0	0	0	0	0	0	0	0	0	0	0	0	0	0	0	0	0	0	0	0	0	0	0	0	0	0	0	0	0	0	0
15	0	0	0	0	0	0	0	0	0	0	0	0	0	0	0	0	0	0	0	0	0	0	0	0	0	0	0	0	0	0	0	0	0

The Caviar Network (Phase 5)

	1	4	89	83	3	5	88	85	86	2	7	6	8	55	47	76	9	34	11	32	84	13	31	12	108	100	18	17	25	82	19	15
1	0	3	2	2	26	3	0	18	0	3	0	3	1	3	1	7	0	1	0	0	0	0	1	2	0	0	0	0	0	1	1	2
4	0	0	0	0	0	0	0	0	0	0	0	0	0	0	0	0	0	0	0	0	0	0	0	0	0	0	0	0	0	0	0	0
89	0	0	0	0	0	0	0	0	0	0	0	0	0	0	0	0	0	0	0	0	0	0	0	0	0	0	0	0	0	0	0	0
83	0	0	0	0	0	0	0	0	2	0	0	0	0	0	0	0	0	0	0	0	0	0	0	0	0	0	0	0	0	0	0	0
3	1	0	0	0	0	0	0	0	0	0	0	0	0	0	0	0	2	0	0	0	0	0	4	4	0	0	0	0	0	0	0	0
5	0	0	0	0	0	0	0	0	0	0	0	0	0	0	0	0	0	0	0	0	0	0	0	0	0	0	0	0	0	0	0	0
88	0	0	0	0	0	0	0	1	0	0	0	0	0	0	0	0	0	0	0	0	0	0	0	0	0	0	0	0	0	0	0	0
85	0	0	0	0	1	0	2	0	0	0	0	0	0	0	0	0	0	0	0	0	0	0	0	0	0	0	0	0	0	0	0	0
86	0	0	0	2	0	0	0	0	0	0	0	0	0	0	0	0	0	0	0	0	0	0	0	0	0	0	0	0	0	0	0	0
2	1	0	0	0	0	0	0	0	0	0	0	0	0	0	0	0	0	0	0	0	0	0	0	0	0	0	0	0	0	0	0	0
7	0	0	1	0	0	0	0	0	0	0	0	0	0	0	0	0	0	0	0	0	0	0	0	0	0	0	0	0	0	0	0	0
6	10	0	0	0	0	0	0	0	0	0	0	0	0	0	0	0	0	0	0	0	0	0	0	0	0	0	0	0	0	0	0	0
8	0	0	0	0	0	1	0	0	0	0	0	0	0	0	0	0	0	0	0	0	0	0	0	0	0	0	0	0	0	0	0	0
55	0	0	0	0	0	0	0	0	0	0	0	0	0	0	0	0	0	0	0	0	0	0	0	0	0	0	0	0	0	0	0	0
47	0	0	0	0	0	0	0	0	0	0	0	0	0	0	0	0	0	0	0	0	0	0	0	0	0	0	0	0	0	0	0	0
76	2	0	1	0	0	0	0	0	0	0	0	0	0	0	0	0	0	0	0	0	0	0	0	0	0	0	0	0	0	0	0	0
9	0	0	0	0	0	0	0	0	0	0	0	0	0	0	0	0	0	0	0	0	0	0	0	0	0	0	0	0	0	0	0	0
34	0	0	0	0	0	0	0	0	0	0	0	0	0	0	0	0	0	0	0	0	0	0	0	0	0	0	0	0	0	0	0	0
11	19	0	0	0	0	0	0	0	0	0	0	0	0	0	0	0	0	0	0	0	0	0	0	0	0	0	0	0	0	0	0	0
32	0	0	0	0	0	0	0	0	0	0	0	0	0	0	0	0	0	0	0	0	0	0	0	1	0	0	0	0	0	0	0	0
84	2	0	0	0	0	0	0	0	0	0	0	0	0	0	0	0	0	0	0	0	0	0	0	0	0	0	0	0	0	0	0	0
13	0	0	0	0	0	0	0	0	0	0	0	0	0	0	0	0	0	0	0	0	0	0	0	0	0	0	0	0	0	0	0	0
31	3	0	0	0	0	0	0	0	0	0	0	0	0	0	0	0	0	0	0	0	0	1	0	0	0	0	0	0	0	0	0	0
12	0	0	0	0	1	0	0	0	0	0	0	0	0	0	0	0	2	0	0	0	0	0	3	0	0	0	1	0	1	0	0	0
108	4	0	0	0	0	0	0	0	0	0	0	1	0	0	0	0	0	0	0	0	0	0	0	0	0	0	0	0	0	0	0	0
100	1	0	0	0	0	0	0	0	0	0	0	0	0	0	0	0	0	0	0	0	0	0	0	0	0	0	0	0	0	0	0	0
18	0	0	0	0	0	0	0	0	0	0	0	0	0	0	0	0	0	0	0	0	0	0	0	3	0	0	0	0	0	0	0	0
17	0	0	0	0	0	0	0	0	0	0	0	0	0	0	0	0	0	0	0	0	0	0	0	2	0	0	0	0	0	0	0	0
25	0	0	0	0	0	0	0	0	0	0	0	0	0	0	0	0	0	0	0	0	0	0	0	0	0	0	0	0	0	0	0	0
82	1	0	0	0	0	0	0	0	0	0	0	0	0	0	0	0	0	0	0	0	0	0	0	0	0	0	0	0	0	0	0	0
19	1	0	0	0	0	0	0	0	0	0	0	0	0	0	0	0	0	0	0	0	0	0	0	0	0	0	0	0	0	0	0	0
15	0	0	0	0	0	0	0	0	0	0	0	0	0	0	0	0	0	0	0	0	0	0	0	0	0	0	0	0	0	0	0	0

The Caviar Network (Phase 6)

	1	4	83	3	5	85	2	6	8	76	9	11	84	13	31	12	14	18	17	25	82	19	78	77	87	20	15
1	0	1	12	29	14	21	8	0	12	22	0	0	0	0	0	2	0	0	0	0	1	11	0	2	0	0	2
4	0	0	0	0	0	0	0	0	0	0	0	0	0	0	0	0	0	0	0	0	0	0	0	0	0	0	0
83	1	0	0	0	0	0	0	0	0	0	0	0	0	0	0	0	0	0	0	0	0	0	0	0	0	0	0
3	62	0	1	0	0	1	0	0	2	8	3	0	0	0	0	17	1	0	0	0	0	0	0	1	0	0	0
5	61	0	0	3	0	0	0	0	2	0	0	0	0	0	0	0	0	0	0	0	0	0	0	0	0	0	0
85	2	0	0	0	0	0	0	0	1	1	0	0	0	0	0	0	0	0	0	0	0	0	0	0	0	0	0
2	10	0	0	2	0	0	0	0	0	0	0	0	0	0	0	0	0	0	0	0	0	0	0	0	0	0	0
6	7	0	0	0	0	0	0	0	0	0	0	0	0	0	0	0	0	0	0	0	0	0	0	0	0	0	0
8	2	0	0	0	0	0	0	0	0	0	0	0	0	0	0	0	0	0	0	0	0	0	0	0	0	0	0
76	42	0	0	4	0	0	0	0	0	0	0	0	0	0	0	0	0	0	0	0	0	1	3	0	0	0	0
9	0	0	0	0	0	0	0	0	0	0	0	0	0	0	0	0	0	0	0	0	0	0	0	0	0	0	0
11	33	0	0	0	0	0	0	0	0	0	0	0	0	0	0	0	0	0	0	0	0	0	0	0	0	8	0
84	2	0	0	0	0	1	0	0	0	0	0	0	0	0	0	0	0	0	0	0	0	0	0	0	0	0	0
13	0	0	0	0	0	0	0	0	0	0	0	0	0	0	0	2	0	0	0	0	0	0	0	0	0	0	0
31	0	0	0	0	0	0	0	0	0	0	0	0	0	0	0	0	0	0	0	0	0	0	0	0	0	0	0
12	2	0	0	32	0	0	0	0	0	5	1	0	0	3	1	0	4	31	25	3	0	0	0	0	0	0	0
14	0	0	0	0	0	0	0	0	0	0	0	0	0	0	0	1	0	0	0	0	0	0	0	0	0	0	0
18	0	0	0	0	0	0	0	0	0	0	0	0	0	0	0	8	0	0	0	0	0	0	0	0	0	0	0
17	0	0	0	0	0	0	0	0	0	0	0	0	0	0	0	15	0	0	0	0	0	0	0	0	0	0	0
25	0	0	0	0	0	0	0	0	0	0	0	0	0	0	0	1	0	0	0	0	0	0	0	0	0	0	0
82	1	0	0	0	0	0	0	0	0	0	0	0	0	0	0	0	0	0	0	0	0	0	0	0	0	0	0
19	75	0	0	1	0	0	0	0	0	0	0	0	0	0	0	0	0	0	0	0	0	0	0	0	0	0	0
78	0	0	0	0	0	0	0	0	0	16	0	0	0	0	0	0	0	0	0	0	0	0	0	0	0	0	0
77	2	0	0	2	0	0	0	0	0	0	0	0	0	0	0	0	0	0	0	0	0	0	0	0	0	0	0
87	3	0	0	1	0	0	0	0	0	0	0	0	0	0	0	0	0	0	0	0	0	0	0	0	0	0	0
20	10	0	0	0	0	0	0	0	0	0	0	0	0	0	0	0	0	0	0	0	0	0	0	0	0	0	0
15	1	0	0	1	0	0	0	0	0	2	0	0	0	0	0	0	0	0	0	0	0	0	0	0	0	0	0

The Caviar Network (Phase 7)

	1	4	83	3	5	88	85	62	9	2	6	8	55	76	9	34	11	12	14	18	17	19	78	77	87	20	22	74	79	69	61	68	16	75	28	81	15
1	0	1	1	23	3	0	2	1	1	5	5	1	0	26	0	0	1	0	2	0	0	4	0	20	2	0	0	0	0	0	0	0	0	0	1	1	0
4	0	0	0	0	0	0	0	0	0	0	0	0	0	0	0	0	0	0	0	0	0	0	0	0	0	0	0	0	0	0	0	0	0	0	0	0	0
83	0	0	0	0	0	0	0	0	1	0	0	0	0	0	0	0	0	0	0	0	0	0	0	0	0	0	0	0	0	0	0	0	0	0	0	0	0
3	65	0	4	0	0	1	4	0	12	1	0	0	0	0	1	0	0	0	0	0	0	1	0	2	0	0	0	3	0	0	0	0	0	0	0	0	0
5	0	0	0	0	0	0	0	0	0	0	0	0	0	0	0	0	0	0	0	0	0	0	0	0	0	0	0	0	0	0	0	0	0	0	0	0	0
88	0	0	0	0	0	0	0	0	0	0	0	0	0	0	0	0	0	0	0	0	0	0	0	0	0	0	0	0	0	0	0	0	0	0	0	0	0
85	4	0	0	0	0	1	0	0	0	0	0	0	0	0	0	0	0	0	0	0	0	0	0	0	0	0	0	0	0	0	0	0	0	0	0	0	0
62	0	0	0	0	0	0	0	0	0	0	0	0	0	0	0	0	0	0	0	0	0	0	0	0	0	0	0	0	0	0	0	0	0	0	0	0	0
9	0	0	5	0	0	0	0	0	0	0	0	0	0	0	0	0	0	0	0	0	0	0	0	0	0	0	0	0	0	0	0	0	0	0	0	0	0
2	6	0	0	0	0	0	0	0	0	0	0	0	0	0	0	0	0	0	0	0	0	0	0	0	0	0	0	0	0	0	0	0	0	0	0	0	0
6	4	0	0	0	0	0	0	0	0	0	0	0	0	0	0	0	0	0	0	0	0	0	0	0	0	0	0	0	0	0	0	0	0	0	0	0	0
8	0	0	0	0	0	0	0	0	0	0	0	0	0	0	0	0	0	0	0	0	0	0	0	0	0	0	0	0	0	0	0	0	0	0	0	0	0
55	1	0	0	0	0	0	0	0	0	0	0	0	0	0	0	0	0	0	0	0	0	0	0	0	0	0	0	0	0	0	0	0	0	0	0	0	0
76	4	0	0	0	0	0	0	0	0	0	0	0	0	0	0	0	0	0	0	0	0	1	9	0	0	0	0	0	1	0	2	0	0	0	0	0	0
9	0	0	0	0	0	0	0	0	0	0	0	0	0	0	0	0	0	0	0	0	0	0	0	0	0	0	0	0	0	0	0	0	0	0	0	0	0
34	1	0	0	0	0	0	0	0	0	0	0	0	0	0	0	0	0	0	0	0	0	0	0	0	0	0	0	0	0	0	0	0	0	0	0	0	0
11	29	0	0	0	0	0	0	0	0	0	0	0	0	0	0	0	0	0	0	0	0	0	0	0	0	8	0	0	0	0	0	0	0	0	0	0	0
12	0	0	0	0	0	0	0	0	0	0	0	0	0	0	0	0	0	0	0	0	0	0	0	0	0	0	1	0	0	0	0	0	0	0	0	0	0
14	0	0	0	0	0	0	0	0	0	0	0	0	0	0	0	0	0	0	0	0	0	0	0	0	0	0	0	0	0	0	0	0	0	0	0	0	0
18	0	0	0	0	0	0	0	0	0	0	0	0	0	0	0	0	0	6	0	0	0	0	0	0	0	0	0	0	0	0	0	0	0	0	0	0	0
17	0	0	0	0	0	0	0	0	0	0	0	0	0	0	0	0	0	1	0	0	0	0	0	0	0	0	0	0	0	0	0	0	0	0	0	0	0
19	21	0	0	0	0	0	0	0	0	0	0	0	0	1	0	0	0	0	0	0	0	0	0	0	0	0	0	0	0	0	0	0	0	0	0	0	0
78	0	0	0	0	0	0	0	0	0	0	0	0	0	7	0	0	0	0	0	0	0	0	0	0	0	0	0	0	1	0	0	0	0	0	0	0	0
77	13	0	0	0	0	0	0	0	0	0	0	0	0	0	0	0	0	0	0	0	0	1	0	0	0	0	0	0	0	0	0	0	0	0	0	0	0
87	15	0	0	2	0	0	0	0	0	0	0	0	0	0	0	0	0	0	0	0	0	0	0	0	0	0	0	0	0	0	0	0	0	0	0	2	0
20	5	0	0	0	0	0	0	0	0	0	0	0	0	0	0	0	0	0	0	0	0	0	0	0	0	0	0	0	0	0	0	0	0	0	0	0	0
22	0	0	0	0	0	0	0	0	0	0	0	0	0	0	0	0	0	1	0	0	0	0	0	0	0	0	0	0	0	0	0	0	0	0	0	0	0
74	0	0	0	0	0	0	1	0	0	0	0	0	0	0	0	0	0	0	0	0	0	0	0	0	0	0	0	0	0	0	0	0	0	0	0	0	0
79	0	0	0	0	0	0	0	0	0	0	0	0	0	0	0	0	0	0	0	0	0	0	3	0	0	0	0	0	0	0	0	0	0	0	0	0	0
69	11	0	0	0	0	0	0	0	0	0	0	0	0	0	0	0	0	0	0	0	0	0	0	0	0	0	0	0	0	0	0	0	0	0	0	0	0
61	0	0	0	0	0	0	0	0	0	0	0	0	0	0	0	0	0	0	0	0	0	0	0	0	0	0	0	0	0	0	0	0	0	0	0	0	0
68	11	0	0	0	0	0	0	0	0	0	0	0	0	0	0	0	0	0	0	0	0	0	0	0	0	0	0	0	0	0	0	0	0	0	0	0	0
16	0	0	0	0	0	0	0	0	0	0	0	0	0	0	0	0	0	4	0	0	0	0	0	0	0	0	0	0	0	0	0	0	0	0	0	0	0
75	0	0	0	0	0	0	0	0	0	0	0	0	0	0	0	0	0	1	0	0	0	0	0	0	0	0	0	0	0	0	0	0	0	0	0	0	0
28	0	0	0	0	0	0	0	0	0	0	0	0	0	0	0	0	0	0	0	0	0	0	0	0	0	0	0	0	0	0	0	0	0	0	0	0	0
81	0	0	0	0	0	0	0	0	0	0	0	0	0	0	0	0	0	0	0	0	0	0	0	0	0	0	0	0	0	0	0	0	0	0	0	0	0
15	1	0	0	0	0	0	0	0	0	0	0	0	0	0	0	0	0	0	0	0	0	0	0	0	0	0	0	0	0	0	0	0	0	0	0	0	0

The Caviar Network (Phase 8)

	1	4	83	3	85	86	2	6	8	76	9	34	35	11	84	13	12	14	18	17	25	82	19	78	77	87	20	22	16	28	81	67	38	73	39	59	37	23	33	80	91	36
1	0	2	2	76	12	0	4	0	2	1	1	0	0	7	1	0	0	1	0	0	0	8	3	0	6	22	6	0	0	3	2	3	0	0	0	0	0	0	0	0	0	0
4	1	0	0	0	0	0	0	0	0	0	0	0	0	0	0	0	0	0	0	0	0	0	0	0	0	0	0	0	0	0	0	0	0	0	0	0	0	0	0	0	0	0
83	0	0	0	0	0	0	0	0	0	0	0	0	0	0	0	0	0	0	0	0	0	0	0	0	0	0	0	0	0	0	0	0	0	0	0	0	0	0	0	0	0	0
3	6	0	0	0	4	1	0	0	6	0	0	2	0	0	0	0	0	1	0	0	0	1	0	0	0	3	0	0	0	0	0	0	0	0	0	0	9	0	0	0	0	0
85	1	0	0	7	0	0	0	0	1	0	0	0	0	0	0	0	0	0	0	0	0	0	0	0	0	0	0	0	0	0	0	0	0	0	0	0	0	0	0	0	0	0
86	0	0	0	0	0	0	0	0	0	0	0	0	0	0	0	0	0	0	0	0	0	0	0	0	0	0	1	0	0	0	0	0	0	0	0	0	0	0	0	0	0	0
2	2	0	0	1	0	0	0	0	0	0	0	1	1	0	0	0	0	0	0	0	0	0	0	0	0	0	0	0	0	0	0	0	0	0	0	0	0	0	0	0	0	0
6	1	0	0	0	0	0	0	0	0	0	0	0	0	0	0	0	0	0	0	0	0	0	0	0	0	0	0	0	0	0	0	0	0	0	0	0	0	0	0	0	0	0
8	0	0	0	0	0	0	0	0	0	0	0	0	0	0	0	0	0	0	0	0	0	0	0	0	0	0	0	0	0	0	0	0	0	0	0	0	0	0	0	0	0	0
76	1	0	0	1	0	0	0	0	0	0	0	0	0	0	0	0	0	0	0	0	0	0	3	0	0	0	0	0	0	0	0	0	0	1	0	3	0	0	0	0	0	0
9	0	0	0	0	0	0	0	0	0	0	0	0	0	0	0	0	0	0	0	0	0	0	0	0	0	0	0	0	0	0	0	0	0	0	0	0	0	0	0	0	0	0
34	0	0	0	0	0	0	1	0	0	0	0	0	0	0	0	0	0	0	0	0	0	0	0	0	0	0	0	0	0	0	0	0	0	0	0	0	0	0	0	0	0	0
35	0	0	0	0	0	0	0	0	0	0	0	0	0	0	0	0	0	0	0	0	0	0	0	0	0	0	0	0	0	0	0	0	0	0	0	0	0	0	0	0	0	0
11	0	0	0	1	0	0	0	0	0	0	0	0	0	0	0	0	0	0	0	0	0	0	0	0	0	0	0	0	0	0	0	0	0	0	0	0	0	0	0	0	0	3
84	0	0	0	0	0	0	0	0	0	0	0	0	0	0	0	0	0	0	0	0	0	0	0	0	0	0	0	0	0	0	0	0	0	0	0	0	0	0	0	0	0	0
13	0	0	0	0	0	0	0	0	0	0	0	0	0	0	0	0	0	0	0	0	0	0	0	0	0	0	0	0	0	0	0	0	0	0	0	0	0	0	0	0	0	0
12	0	0	0	0	0	0	0	0	0	0	0	0	0	0	0	1	0	16	0	0	0	0	0	0	0	0	0	3	0	0	0	0	0	0	0	0	0	1	0	0	0	0
14	1	0	0	1	0	0	0	0	0	0	0	0	0	0	0	0	1	0	0	0	0	0	0	0	0	0	0	0	0	0	0	0	0	0	0	0	0	0	0	0	0	0
18	0	0	0	0	0	0	0	0	0	0	0	0	0	0	0	0	6	0	0	0	0	0	0	0	0	0	0	0	0	0	0	0	0	0	0	0	0	0	0	0	0	0
17	0	0	0	0	0	0	0	0	0	0	0	0	0	0	0	0	6	0	0	0	0	0	0	0	0	0	0	0	0	0	0	0	0	0	0	0	0	0	0	0	0	0
25	0	0	0	0	0	0	0	0	0	0	0	0	0	0	0	0	3	0	0	0	0	0	0	0	0	0	0	0	0	0	0	0	0	0	0	0	0	0	0	0	0	0
82	0	0	0	0	0	0	0	0	0	0	0	0	0	0	0	0	0	0	0	0	0	0	0	0	0	0	0	0	0	0	0	0	0	0	0	0	0	0	0	0	0	0
19	1	0	0	0	0	0	0	0	0	0	0	0	0	0	0	0	0	0	0	0	0	0	0	0	0	0	0	0	0	0	0	0	0	0	0	0	0	0	0	0	0	0
78	0	0	0	0	0	0	0	0	0	1	0	0	0	0	0	0	0	0	0	0	0	0	0	0	0	0	0	0	0	0	0	0	0	0	0	0	0	0	0	0	0	0
77	0	0	0	0	0	0	0	0	0	0	0	0	0	0	0	0	0	0	0	0	0	0	0	0	0	0	0	0	0	0	0	0	0	0	0	0	0	0	0	0	0	0
87	6	0	0	1	0	0	0	0	0	0	1	0	0	0	0	0	0	0	0	0	0	0	0	0	0	0	0	0	0	0	4	0	1	1	8	0	4	0	0	0	3	0
20	0	0	0	0	0	0	0	0	0	0	0	0	0	0	0	0	0	0	0	0	0	0	0	0	0	0	0	0	0	0	0	0	0	0	0	0	0	0	0	0	0	0
22	0	0	0	3	0	0	0	0	0	0	0	0	0	0	0	0	0	0	0	0	0	0	0	0	0	0	0	0	0	0	0	0	0	0	0	0	0	0	0	0	0	0
16	0	0	0	0	0	0	0	0	0	0	0	0	0	0	0	0	25	0	0	0	0	0	0	0	0	0	0	0	0	0	0	0	0	0	0	0	0	0	0	0	0	0
28	1	0	0	0	0	0	0	0	0	0	0	0	0	0	0	0	0	0	0	0	0	0	0	0	0	0	0	0	0	0	0	0	0	0	0	0	0	0	0	0	0	0
81	0	0	0	0	0	0	0	0	0	0	0	0	0	0	0	0	0	0	0	0	0	0	0	0	0	0	0	0	0	0	0	0	0	0	0	0	0	0	0	0	0	0
67	0	0	0	0	0	0	0	0	0	0	0	0	0	0	0	0	0	0	0	0	0	0	0	0	0	0	0	0	0	0	0	0	0	0	0	0	0	0	0	0	0	0
38	0	0	0	0	0	0	0	0	0	0	0	0	0	0	0	0	0	0	0	0	0	0	0	0	0	1	0	0	0	0	0	0	0	0	0	0	0	0	0	0	0	0
73	0	0	0	0	0	0	0	0	0	0	0	0	0	0	0	0	0	0	0	0	0	0	0	0	0	0	0	0	0	0	0	0	0	0	0	0	0	0	0	0	0	0
39	0	0	0	0	0	0	0	0	0	0	0	0	0	0	0	0	0	0	0	0	0	0	0	0	0	0	0	0	0	0	0	0	0	0	0	0	0	0	0	0	0	0
59	0	0	0	0	0	0	0	0	0	0	0	0	0	0	0	0	0	0	0	0	0	0	0	0	0	0	0	0	0	0	0	0	0	0	0	0	0	0	0	0	0	0
37	0	0	0	0	0	0	0	0	0	0	0	0	0	0	0	0	0	0	0	0	0	0	0	0	0	2	0	0	0	0	0	0	0	0	0	0	0	0	0	0	0	0
23	0	0	0	0	0	0	0	0	0	0	0	0	0	0	0	0	5	0	0	0	0	0	0	0	0	0	0	0	0	0	0	0	0	0	0	0	0	0	0	0	0	0
33	0	0	0	0	0	0	0	0	0	0	0	0	0	0	0	0	1	0	0	0	0	0	0	0	0	0	0	0	0	0	0	0	0	0	0	0	0	0	0	0	0	0
80	0	0	0	0	0	0	0	0	0	0	0	0	0	0	0	0	1	0	0	0	0	0	0	0	0	0	0	0	0	0	0	0	0	0	0	0	0	0	0	0	0	0
91	0	0	0	0	0	0	0	0	0	0	0	0	0	0	0	0	0	0	0	0	0	0	0	0	0	0	0	0	0	0	0	0	0	0	0	0	0	0	0	0	0	0
36	0	0	0	0	0	0	0	0	0	0	0	0	0	0	0	0	0	0	0	0	0	0	0	0	0	0	0	0	0	0	0	0	0	0	0	0	0	0	0	0	0	0

The Caviar Network (Phase 9)

	1	89	83	3	88	85	90	2	7	6	8	76	11	13	12	14	18	17	82	78	87	79	16	81	59	37	36	96	46	29	30	105	101	41
1	0	0	3	0	0	6	0	0	0	1	1	7	1	0	0	0	0	0	42	0	12	0	0	0	0	0	0	1	0	0	0	0	0	0
89	0	0	0	0	0	0	0	0	0	0	0	0	0	0	0	0	0	0	0	0	0	0	0	0	0	0	0	0	0	0	0	0	0	0
83	0	0	0	0	0	0	0	0	0	0	0	0	0	0	0	0	0	0	0	0	0	0	0	0	0	0	0	0	0	0	0	0	0	0
3	1	1	0	0	0	11	1	1	1	0	11	3	0	0	1	0	0	0	2	0	17	0	0	0	0	0	0	0	0	0	0	0	0	0
88	0	0	0	0	0	1	0	0	0	0	0	0	0	0	0	0	0	0	0	0	0	0	0	0	0	0	0	0	0	0	0	0	0	0
85	4	0	0	11	0	0	0	0	0	0	0	0	0	0	0	0	0	0	0	0	0	0	0	0	0	0	0	0	0	0	0	0	0	0
90	0	0	0	0	0	0	0	0	0	0	0	0	0	0	0	0	0	0	0	0	0	0	0	0	0	0	0	0	0	0	0	0	0	0
2	0	0	0	0	0	0	0	0	0	0	0	0	0	0	0	0	0	0	0	0	0	0	0	0	0	0	0	0	0	0	0	0	0	0
7	0	0	0	0	0	0	0	0	0	0	0	0	0	0	0	0	0	0	0	0	0	1	0	0	0	0	0	0	0	0	0	0	0	0
6	0	0	0	0	0	0	0	0	0	0	0	0	0	0	0	0	0	0	0	0	0	0	0	0	0	0	0	0	0	0	0	0	0	0
8	2	0	0	1	0	0	0	0	0	0	0	0	0	0	0	0	0	0	0	0	0	0	0	0	0	0	0	0	0	0	0	0	0	0
76	0	0	0	1	0	0	0	0	0	0	0	0	0	0	0	0	0	0	0	0	1	0	0	0	1	0	0	0	0	0	0	0	0	0
11	0	0	0	0	0	0	0	0	0	0	0	0	0	0	0	0	0	0	0	0	0	0	0	0	0	0	0	0	0	0	0	0	0	0
13	0	0	0	0	0	0	0	0	0	0	0	0	0	0	0	0	0	0	0	0	0	0	0	0	0	0	0	0	0	0	0	0	0	0
12	0	0	0	1	0	0	0	0	0	0	0	0	0	2	0	13	0	4	0	0	0	0	9	0	0	0	0	0	0	6	1	0	0	0
14	0	0	0	0	0	0	0	0	0	0	0	0	0	0	0	0	0	0	0	0	0	0	0	0	0	0	0	0	0	0	0	0	0	0
18	0	0	0	0	0	0	0	0	0	0	0	0	0	0	6	0	0	0	0	0	0	0	0	0	0	0	0	0	0	0	0	0	0	0
17	0	0	0	0	0	0	0	0	0	0	0	0	0	0	2	0	0	0	0	0	0	0	0	0	0	0	0	0	0	0	0	0	0	0
82	10	0	0	0	0	0	0	0	0	0	0	0	0	0	0	0	0	0	0	0	1	0	0	0	0	0	1	4	22	0	0	0	0	0
78	0	0	0	0	0	0	0	0	0	0	0	1	0	0	0	0	0	0	0	0	0	0	0	0	0	0	0	0	0	0	0	0	0	0
87	4	0	0	25	0	0	0	0	0	0	0	0	0	0	0	0	0	0	8	0	0	0	0	5	0	10	0	0	0	0	0	2	0	3
79	0	0	0	0	0	0	0	0	0	0	0	0	0	0	0	0	0	0	0	1	0	0	0	0	0	0	0	0	0	0	0	0	1	0
16	0	0	0	0	0	0	0	0	0	0	0	0	0	0	12	0	0	0	0	0	0	0	0	0	0	0	0	0	0	0	0	0	0	0
81	0	0	0	0	0	0	0	0	0	0	0	0	0	0	0	0	0	0	0	0	1	0	0	0	0	0	0	0	0	0	0	0	0	0
59	0	0	0	0	0	0	0	0	0	0	0	0	0	0	0	0	0	0	0	0	0	0	0	0	0	0	0	0	0	0	0	0	0	0
37	0	0	0	0	0	0	0	0	0	0	0	0	0	0	0	0	0	0	0	0	0	0	0	0	0	0	0	0	0	0	0	0	0	0
36	0	0	0	0	0	0	0	0	0	0	0	0	0	0	0	0	0	0	0	0	0	0	0	0	0	0	0	0	0	0	0	0	0	0
96	0	0	0	0	0	0	0	0	0	0	0	0	0	0	0	0	0	0	0	0	0	0	0	0	0	0	0	0	4	0	1	0	0	0
46	0	0	0	0	0	0	0	0	0	0	0	0	0	0	0	0	0	0	5	0	0	0	0	0	0	0	0	4	0	0	0	0	0	0
29	0	0	0	0	0	0	0	0	0	0	0	0	0	0	1	0	0	0	0	0	0	0	0	0	0	0	0	0	0	0	0	0	0	0
30	0	0	0	0	0	0	0	0	0	0	0	0	0	0	0	0	0	0	0	0	0	0	0	0	0	0	0	0	0	0	0	0	0	0
105	0	0	0	0	0	0	0	0	0	0	0	0	0	0	0	0	0	0	0	0	0	0	0	0	0	0	0	0	0	0	0	0	0	0
101	0	0	0	0	0	0	0	0	0	0	0	0	0	0	0	0	0	0	0	0	0	0	0	0	0	0	0	0	0	0	0	0	0	0
41	0	0	0	0	0	0	0	0	0	0	0	0	0	0	0	0	0	0	0	0	0	0	0	0	0	0	0	0	0	0	0	0	0	0

The Caviar Network (Phase 10)

	1	4	83	3	85	86	8	76	9	84	12	14	18	17	82	19	87	22	16	81	38	73	37	24	58	96	46	105	41	21	103	104	27	95	93	42	71	40	44	45	65	70
1	0	1	4	3	6	0	7	3	0	1	0	0	0	0	17	0	24	0	0	4	0	0	2	0	0	0	0	1	0	0	0	0	0	0	0	0	0	0	0	0	0	5
4	0	0	0	0	0	0	0	0	0	0	0	0	0	0	0	0	0	0	0	0	0	0	0	0	0	0	0	0	0	0	0	0	0	0	0	0	0	0	0	0	0	0
83	1	0	0	0	0	4	0	0	0	0	0	0	0	0	0	0	0	0	0	0	0	0	0	0	0	0	0	0	0	0	0	0	0	0	0	0	0	0	0	0	0	0
3	2	0	0	0	0	0	0	0	0	0	0	0	0	0	0	0	0	0	0	0	0	0	0	0	0	0	0	0	0	0	0	0	0	0	0	0	0	0	0	0	0	0
85	11	0	0	0	0	0	0	1	0	0	0	0	0	0	0	0	0	0	0	0	0	0	0	0	0	0	0	0	0	0	0	0	0	0	0	0	0	0	0	0	0	0
86	0	0	0	0	0	0	0	0	0	0	0	0	0	0	0	0	0	0	0	0	0	0	0	0	0	0	0	0	0	0	0	0	0	0	0	0	0	0	0	0	0	0
8	1	0	0	0	0	0	0	0	3	0	0	0	0	0	0	0	0	0	0	0	0	0	0	0	0	0	0	0	0	0	0	0	0	0	0	0	0	0	0	0	0	0
76	7	0	0	0	0	0	0	0	0	0	0	0	0	0	0	0	0	0	0	0	0	0	0	0	0	0	0	0	0	0	0	0	0	0	0	0	0	0	0	0	0	0
9	0	0	0	0	0	0	0	0	0	0	0	0	0	0	0	0	0	0	0	0	0	0	0	0	0	0	0	0	0	0	0	0	0	0	0	0	0	0	0	0	0	0
84	0	0	0	0	14	0	0	0	0	0	0	0	0	0	0	0	0	0	0	0	0	0	0	0	0	0	0	0	0	0	0	0	0	0	0	0	0	0	0	0	0	0
12	0	0	0	0	0	0	0	0	0	0	0	57	2	1	0	0	0	1	34	0	0	0	0	1	0	0	0	0	0	0	0	0	0	0	0	0	0	0	0	0	0	0
14	0	0	0	0	0	0	0	0	0	0	6	0	0	0	0	0	0	0	0	0	0	0	0	1	22	0	0	0	0	0	0	0	0	0	0	0	0	0	0	0	0	0
18	0	0	0	0	0	0	0	0	0	0	0	0	0	0	0	0	0	0	0	0	0	0	0	0	0	0	0	0	0	0	0	0	0	0	0	0	0	0	0	0	0	0
17	0	0	0	0	0	0	0	0	0	0	0	0	0	0	0	0	0	0	0	0	0	0	0	0	0	0	0	0	0	0	0	0	0	0	0	0	0	0	0	0	0	0
82	3	0	0	0	0	0	0	0	0	0	0	0	0	0	0	0	0	0	0	0	0	0	0	0	0	6	3	0	0	0	0	0	0	0	0	0	0	0	0	0	0	0
19	0	0	0	0	0	0	0	0	0	0	0	0	0	0	0	0	0	0	0	0	0	0	0	0	0	0	0	0	0	0	0	0	0	0	0	0	1	0	0	0	0	0
87	14	0	0	0	0	0	0	0	0	0	0	0	0	0	9	0	0	0	0	8	1	1	16	0	0	0	0	3	3	8	0	0	1	0	0	0	0	1	0	0	0	0
22	0	0	0	0	0	0	0	0	0	0	0	0	0	0	0	0	0	0	0	0	0	0	0	0	0	0	0	0	0	0	0	0	0	0	0	0	0	0	0	0	0	0
16	0	0	0	0	0	0	0	0	0	0	8	0	0	0	0	0	0	0	0	0	0	0	0	0	0	0	0	0	0	0	0	0	0	0	0	0	0	0	0	0	0	0
81	0	0	0	0	0	0	0	0	0	0	0	0	0	0	0	0	1	0	0	0	0	0	0	0	0	0	0	0	0	0	0	0	0	0	0	0	0	0	0	0	0	0
38	0	0	0	0	0	0	0	0	0	0	0	0	0	0	0	0	0	0	0	0	0	0	0	0	0	0	0	0	0	0	0	0	0	0	0	0	0	0	0	0	0	0
73	0	0	0	0	0	0	0	0	0	0	0	0	0	0	0	0	0	0	0	0	0	0	0	0	0	0	0	0	0	0	0	0	0	0	0	0	0	0	0	0	0	0
37	1	0	0	0	0	0	0	0	0	0	0	0	0	0	10	0	11	0	0	0	0	0	0	0	0	0	0	0	27	0	0	0	0	21	13	4	0	0	22	0	0	0
24	0	0	0	0	0	0	0	0	0	0	0	1	0	0	0	0	0	0	0	0	0	0	0	0	0	0	0	0	0	0	0	0	0	0	0	0	0	0	0	0	0	0
58	0	0	0	0	0	0	0	0	0	0	0	7	0	0	0	0	0	0	0	0	0	0	0	3	0	0	0	0	0	0	0	0	0	0	0	0	0	0	0	0	0	0
96	0	0	0	0	0	0	0	0	0	0	0	0	0	0	1	0	0	0	0	0	0	0	0	0	0	0	0	0	0	0	0	0	0	0	0	0	0	0	0	0	0	0
46	0	0	0	0	0	0	0	0	0	0	0	0	0	0	1	0	0	0	0	0	0	0	0	0	0	1	0	0	0	0	0	0	0	0	0	0	0	0	0	0	0	0
105	0	0	0	0	0	0	0	0	0	0	0	0	0	0	0	0	0	0	0	0	0	0	0	0	0	0	0	0	0	0	0	0	0	0	0	0	0	0	0	0	0	0
41	0	0	0	0	0	0	0	0	0	0	0	0	0	0	0	0	0	0	0	0	0	0	7	0	0	0	0	0	0	0	0	0	0	0	0	0	0	0	0	0	0	0
21	0	0	0	0	0	0	0	0	0	0	0	0	0	0	0	0	0	0	0	0	0	0	0	0	0	0	0	0	0	0	0	0	0	0	0	0	0	0	0	0	0	0
103	0	0	0	0	0	0	0	0	0	0	0	0	0	0	0	0	0	0	0	0	0	0	0	0	0	0	0	0	0	0	0	0	0	0	0	0	0	0	0	0	0	0
104	0	0	0	0	0	0	0	0	0	0	0	0	0	0	0	0	0	0	0	0	0	0	0	0	0	0	0	0	0	0	1	0	0	0	0	0	0	0	0	0	0	0
27	0	0	0	0	0	0	0	0	0	0	0	0	0	0	0	0	0	0	0	0	0	0	0	0	0	0	0	0	0	0	0	0	0	0	0	0	0	0	0	0	0	0
95	0	0	0	0	0	0	0	0	0	0	0	0	0	0	0	0	0	0	0	0	0	0	0	0	0	0	0	0	0	0	0	0	0	0	0	0	0	0	0	0	0	0
93	0	0	0	0	0	0	0	0	0	0	0	0	0	0	0	0	0	0	0	0	0	0	0	0	0	0	0	0	0	0	0	0	0	0	0	0	0	0	0	0	0	0
42	0	0	0	0	0	0	0	0	0	0	0	0	0	0	0	0	0	0	0	0	0	0	0	0	0	0	0	0	0	0	0	0	0	0	0	0	0	0	0	0	0	0
71	0	0	0	0	0	0	0	3	0	0	0	0	0	0	0	0	0	0	0	0	0	0	0	0	0	0	0	0	0	0	0	0	0	0	0	0	0	0	0	0	0	0
40	0	0	0	0	0	0	0	0	0	0	0	0	0	0	0	0	0	0	0	0	0	0	0	0	0	0	0	0	0	0	0	0	0	0	0	0	0	0	0	0	0	0
44	0	0	0	0	0	0	0	0	0	0	0	0	0	0	0	0	0	0	0	0	0	0	2	0	0	0	0	0	0	0	0	0	0	0	0	0	0	0	0	0	0	0
45	0	0	0	0	0	0	0	0	0	0	0	0	0	0	0	0	0	0	0	0	0	0	7	0	0	0	0	0	0	0	0	0	0	0	0	0	0	0	0	0	0	0
65	0	0	0	0	0	0	0	0	0	0	1	0	0	0	0	0	0	0	0	0	0	0	0	0	0	0	0	0	0	0	0	0	0	0	0	0	0	0	0	0	0	0
70	0	0	0	0	0	0	0	0	0	0	0	0	0	0	0	0	0	0	0	0	0	0	0	0	0	0	0	0	0	0	0	0	0	0	0	0	0	0	0	0	0	0

The Caviar Network (Phase 11)

	1	83	3	88	85	86	76	11	84	13	12	14	18	17	82	78	87	79	61	16	81	59	37	36	24	58	96	46	101	41	27	93	42	43	102	72	26	94	92	66	65
1	0	0	0	0	6	0	5	1	0	0	0	0	0	0	5	0	3	0	0	0	0	0	0	0	0	0	0	0	0	2	0	0	0	0	0	0	0	0	0	0	0
83	0	0	0	0	0	3	0	0	0	0	0	0	0	0	0	0	0	0	0	0	0	0	0	0	0	0	0	0	0	0	0	0	0	0	0	0	0	0	0	0	0
3	0	0	0	0	0	0	0	0	0	0	3	0	0	0	0	0	0	0	0	0	0	0	0	0	0	0	0	0	0	0	0	0	0	0	0	0	0	0	0	0	0
88	0	0	0	0	0	0	0	0	0	0	0	0	0	0	0	0	0	0	0	0	0	0	0	0	0	0	0	0	0	0	0	0	0	0	0	0	0	0	0	0	0
85	3	0	0	0	0	0	4	0	0	0	0	0	0	0	0	0	0	0	0	0	0	0	0	0	0	0	0	0	0	0	0	0	0	0	0	0	0	0	0	0	0
86	0	0	0	0	0	0	0	0	0	0	0	0	0	0	0	0	0	0	0	0	0	0	0	0	0	0	0	0	0	0	0	0	0	0	0	0	0	0	0	0	0
76	7	0	0	1	6	0	0	0	0	0	0	0	0	0	0	0	2	4	3	0	0	0	0	0	0	0	0	0	0	0	0	0	0	0	0	2	0	0	0	0	0
11	1	0	0	0	0	0	0	0	0	0	0	0	0	0	0	0	0	0	0	0	0	0	0	0	0	0	0	0	0	0	0	0	0	0	0	0	0	0	0	0	0
84	1	0	0	0	1	0	0	0	0	0	0	0	0	0	0	0	0	0	0	0	0	0	0	0	0	0	0	0	0	0	0	0	0	0	0	0	0	0	0	0	0
13	0	0	0	0	0	0	0	0	0	0	0	0	0	0	0	0	0	0	0	0	0	0	0	0	0	0	0	0	0	0	0	0	0	0	0	0	0	0	0	0	0
12	0	0	0	0	0	0	0	0	0	1	0	29	3	2	0	0	0	0	0	5	0	0	0	0	0	0	0	0	0	0	2	0	0	0	0	0	2	0	0	4	1
14	0	0	0	0	0	0	0	0	0	0	5	0	0	0	0	0	0	0	0	0	0	0	0	0	0	4	0	0	0	0	0	0	0	0	0	0	0	0	0	0	0
18	0	0	0	0	0	0	0	0	0	0	2	0	0	0	0	0	0	0	0	0	0	0	0	0	0	0	0	0	0	0	0	0	0	0	0	0	0	0	0	0	0
17	0	0	0	0	0	0	0	0	0	0	11	0	0	0	0	0	0	0	0	0	0	0	0	0	0	0	0	0	0	0	0	0	0	0	0	0	0	0	0	0	0
82	1	0	0	0	0	0	0	0	0	0	0	0	0	0	0	0	3	0	0	0	0	0	0	4	0	0	5	0	0	0	0	0	0	0	0	0	0	0	0	0	0
78	0	0	0	0	0	0	0	0	0	0	0	0	0	0	0	0	0	1	0	0	0	0	0	0	0	0	0	0	0	0	0	0	0	0	0	0	0	0	0	0	0
87	2	0	0	0	0	0	2	0	0	0	0	0	0	0	1	0	0	0	0	0	2	0	0	0	0	0	0	0	0	0	0	0	0	0	0	0	0	0	0	0	0
79	0	0	0	0	0	0	0	0	0	0	0	0	0	0	0	2	0	0	0	0	0	2	0	0	0	0	0	0	0	0	0	0	0	0	1	0	0	0	0	0	0
61	0	0	0	0	0	0	0	0	0	0	0	0	0	0	0	0	0	0	0	0	0	0	0	0	0	0	0	0	0	0	0	0	0	0	0	0	0	0	0	0	0
16	0	0	0	0	0	0	0	0	0	0	33	0	0	0	0	0	0	0	0	0	0	0	0	0	0	0	0	0	0	0	0	0	0	0	0	0	0	0	0	0	0
81	0	0	0	0	0	0	0	0	0	0	0	0	0	0	0	0	0	0	0	0	0	0	0	0	0	0	0	0	0	0	0	0	0	0	0	0	0	0	0	0	0
59	0	0	0	0	0	0	0	0	0	0	0	0	0	0	0	0	0	6	0	0	0	0	0	0	0	0	0	0	0	0	0	0	0	0	0	0	0	0	0	0	0
37	0	0	0	0	0	0	0	0	0	0	0	0	0	0	0	0	0	0	0	0	0	0	0	0	0	0	0	0	0	21	4	17	0	0	0	0	0	0	0	0	0
36	0	0	0	0	0	0	0	0	0	0	0	0	0	0	0	0	0	0	0	0	0	0	0	0	0	0	0	0	0	0	0	0	0	0	0	0	0	0	0	0	0
24	0	0	0	0	0	0	0	0	0	0	0	0	0	0	0	0	0	0	0	0	0	0	0	0	0	0	0	0	0	0	0	0	0	0	0	0	0	0	0	0	0
58	0	0	0	0	0	0	0	0	0	0	0	12	0	0	0	0	0	0	0	0	0	0	0	0	3	0	0	0	0	0	0	0	0	0	0	0	0	0	0	0	0
96	0	0	0	0	0	0	0	0	0	0	0	0	0	0	5	0	0	0	0	0	0	0	0	0	0	0	0	2	0	0	0	0	0	0	0	0	0	0	0	0	0
46	0	0	0	0	0	0	0	0	0	0	0	0	0	0	0	0	0	0	0	0	0	0	0	0	0	0	0	0	0	0	0	0	0	0	0	0	0	0	0	0	0
101	0	0	0	0	0	0	0	0	0	0	0	0	0	0	0	0	0	12	0	0	0	0	0	0	0	0	0	0	0	0	0	0	0	0	3	0	0	0	0	0	0
41	0	0	0	0	0	0	0	0	0	12	2	0	0	0	0	0	0	0	0	0	0	0	6	0	0	0	0	0	0	0	4	3	0	0	0	0	0	10	11	0	0
27	0	0	0	0	0	0	0	0	0	0	2	0	0	1	0	0	0	0	0	0	0	0	0	0	0	0	0	0	0	0	0	0	0	0	0	0	0	0	0	0	0
93	0	0	0	0	0	0	0	0	0	0	0	0	0	0	0	0	0	0	0	0	0	0	1	0	0	0	0	0	0	1	0	0	0	0	0	0	0	0	0	0	0
42	0	0	0	0	0	0	0	0	0	0	2	0	0	0	0	0	0	0	0	0	0	0	0	0	0	0	0	0	0	0	0	0	0	0	0	0	0	0	0	0	0
43	0	0	0	0	0	0	0	0	0	0	0	0	0	0	0	0	0	0	0	0	0	0	0	0	0	0	0	0	0	1	0	0	0	0	0	0	0	0	0	0	0
102	0	0	0	0	0	0	0	0	0	0	0	0	0	0	0	0	0	2	0	0	0	1	0	0	0	0	0	0	0	0	0	0	0	0	0	0	0	0	0	0	0
72	0	0	0	0	0	0	0	0	0	0	0	0	0	0	0	0	0	0	0	0	0	0	0	0	0	0	0	0	0	0	0	0	0	0	0	0	0	0	0	0	0
26	0	0	0	0	0	0	0	0	0	0	0	0	0	0	0	0	0	0	0	0	0	0	0	0	0	0	0	0	0	0	0	0	0	0	0	0	0	0	0	0	0
94	0	0	0	0	0	0	0	0	0	0	0	0	0	0	0	0	0	0	0	0	0	0	0	0	0	0	0	0	0	0	0	0	0	0	0	0	0	0	0	0	0
92	0	0	0	0	0	0	0	0	0	0	0	0	0	0	0	0	0	0	0	0	0	0	0	0	0	0	0	0	0	0	0	0	0	0	0	0	0	0	0	0	0
66	0	0	0	0	0	0	0	0	0	0	0	0	0	0	0	0	0	0	0	0	0	0	0	0	0	0	0	0	0	0	0	0	0	0	0	0	0	0	0	0	0
65	0	0	0	0	0	0	0	0	0	0	18	0	0	0	0	0	0	0	0	0	0	0	0	0	0	0	0	0	0	0	0	0	0	0	0	0	0	0	0	0	0

The Siren Network

	1	2	3	4	5	6	7	8	9	10	11	12	13	14	15	16	17	18	19	20	21	22	23	24	25	26	27	28	29	30	31	32	33	34	35	36	37	38	39	40	41	42	43	44
1	0	1	1	1	1	1	1	1	1	1	1	1	1	1	1	1	1	1	1	1	1	1	1	1	1	1	1	0	0	0	0	0	0	0	0	1	1	1	1	0	1	0	1	1
2	1	0	0	0	0	0	0	0	0	0	0	0	0	0	0	0	0	0	0	0	0	0	0	0	0	0	0	0	0	0	0	0	0	0	0	0	0	0	0	0	0	0	0	0
3	1	0	0	0	0	1	0	0	0	0	0	0	0	0	0	0	0	0	0	0	0	0	0	0	0	0	0	0	0	0	0	0	0	0	0	0	0	0	0	0	0	0	0	0
4	1	0	0	0	0	0	0	0	0	0	0	0	0	0	0	0	0	0	0	0	0	0	0	0	0	0	0	0	0	0	0	0	0	0	0	0	0	0	0	0	0	0	0	0
5	1	0	0	0	0	1	0	1	0	0	0	0	0	0	1	0	0	0	0	0	0	0	0	0	0	0	0	0	0	0	0	0	0	0	0	0	0	1	0	0	0	0	1	0
6	1	0	1	0	1	0	0	1	0	1	0	0	0	0	1	0	1	0	0	1	0	0	0	1	0	0	0	0	0	0	0	0	0	0	0	0	0	0	0	0	0	0	1	1
7	1	0	0	0	0	0	0	0	0	0	0	0	0	0	0	0	0	0	0	0	0	0	0	0	0	0	0	0	0	0	0	0	0	0	0	0	0	0	0	0	0	0	0	0
8	1	0	0	0	1	1	0	0	0	1	0	0	0	0	1	0	0	0	0	0	0	0	0	1	0	0	0	0	0	0	0	0	0	0	0	0	0	0	0	0	0	0	0	1
9	1	0	0	0	0	0	0	0	0	0	0	0	0	0	0	0	0	0	0	0	0	0	0	0	0	0	0	0	0	0	0	0	0	0	0	0	0	0	0	0	0	0	0	0
10	1	0	0	0	0	1	0	1	0	0	0	0	0	0	1	0	0	0	0	0	0	0	0	1	0	0	0	0	0	0	0	0	0	0	0	1	0	0	0	0	0	0	0	1
11	1	0	0	0	0	0	0	0	0	0	0	0	0	0	0	0	0	0	0	0	0	0	0	0	0	0	0	0	0	0	0	0	0	0	0	0	0	0	0	1	0	1	0	0
12	1	0	0	0	0	0	0	0	0	0	0	0	0	0	0	0	0	0	0	0	0	0	0	0	0	0	0	0	0	0	0	0	0	0	0	0	0	0	0	0	0	0	0	0
13	1	0	0	0	0	0	0	0	0	0	0	0	0	0	0	0	0	0	0	0	0	0	0	0	0	0	0	0	0	0	0	0	0	0	0	0	0	0	0	0	0	0	0	0
14	1	0	0	0	0	0	0	0	0	0	0	0	0	0	0	0	0	0	0	0	0	0	0	0	0	0	0	0	0	0	0	0	0	0	0	0	0	0	0	0	0	0	0	0
15	1	0	0	0	1	1	0	1	0	1	0	0	0	0	0	0	0	0	0	0	0	0	0	1	0	0	0	1	1	1	1	1	1	1	1	0	0	0	0	0	0	0	0	1
16	1	0	0	0	0	0	0	0	0	0	0	0	0	0	0	0	0	0	0	0	0	0	0	0	0	0	0	0	0	0	0	0	0	0	0	0	0	0	1	0	0	0	0	0
17	1	0	0	0	0	1	0	0	0	0	0	0	0	0	0	0	0	0	0	0	0	0	0	0	0	0	0	0	0	0	0	0	0	0	0	0	0	0	0	0	0	0	0	0
18	1	0	0	0	0	0	0	0	0	0	0	0	0	0	0	0	0	0	0	0	0	0	0	0	0	0	0	0	0	0	0	0	0	0	0	0	0	0	0	0	0	0	0	0
19	1	0	0	0	0	0	0	0	0	0	0	0	0	0	0	0	0	0	0	0	0	0	0	0	0	0	0	0	0	0	0	0	0	0	0	0	0	0	0	0	0	0	0	0
20	1	0	0	0	0	1	0	0	0	0	0	0	0	0	0	0	0	0	0	0	0	0	0	1	0	0	0	0	0	0	0	0	0	0	0	0	0	0	0	0	0	0	0	0
21	1	0	0	0	0	0	0	0	0	0	0	0	0	0	0	0	0	0	0	0	0	0	0	0	0	0	0	0	0	0	0	0	0	0	0	0	0	0	0	0	0	0	0	0
22	1	0	0	0	0	0	0	0	0	0	0	0	0	0	0	0	0	0	0	0	0	0	0	0	0	0	0	0	0	0	0	0	0	0	0	0	0	0	0	0	0	0	0	1
23	1	0	0	0	0	0	0	0	0	0	0	0	0	0	0	0	0	0	0	0	0	0	0	0	0	0	0	0	0	0	0	0	0	0	0	0	0	0	0	0	0	0	0	0
24	1	0	0	0	0	1	0	1	0	1	0	0	0	0	1	0	0	0	0	1	0	0	0	0	1	0	0	0	0	0	0	0	0	0	0	0	0	0	0	0	0	0	0	1
25	1	0	0	0	0	0	0	0	0	0	0	0	0	0	0	0	0	0	0	0	0	0	0	1	0	1	0	0	0	0	0	0	0	0	0	0	0	0	0	0	0	0	0	0
26	1	0	0	0	0	0	0	0	0	0	0	0	0	0	0	0	0	0	0	0	0	0	0	0	1	0	0	0	0	0	0	0	0	0	0	0	0	0	0	0	0	0	0	0
27	1	0	0	0	0	0	0	0	0	0	0	0	0	0	0	0	0	0	0	0	0	0	0	0	0	0	0	0	0	0	0	0	0	0	0	0	0	0	0	0	0	0	0	0
28	0	0	0	0	0	0	0	0	0	0	0	0	0	0	1	0	0	0	0	0	0	0	0	0	0	0	0	0	1	1	1	1	1	1	1	0	0	0	0	0	0	0	0	0
29	0	0	0	0	0	0	0	0	0	0	0	0	0	0	1	0	0	0	0	0	0	0	0	0	0	0	0	1	0	1	1	1	1	1	1	0	0	0	0	0	0	0	0	0
30	0	0	0	0	0	0	0	0	0	0	0	0	0	0	1	0	0	0	0	0	0	0	0	0	0	0	0	1	1	0	1	1	1	1	1	0	0	0	0	0	0	0	0	0
31	0	0	0	0	0	0	0	0	0	0	0	0	0	0	1	0	0	0	0	0	0	0	0	0	0	0	0	1	1	1	0	1	1	1	1	0	0	0	0	0	0	0	0	0
32	0	0	0	0	0	0	0	0	0	0	0	0	0	0	1	0	0	0	0	0	0	0	0	0	0	0	0	1	1	1	1	0	1	1	1	0	0	0	0	0	0	0	0	0
33	0	0	0	0	0	0	0	0	0	0	0	0	0	0	1	0	0	0	0	0	0	0	0	0	0	0	0	1	1	1	1	1	0	1	1	0	0	0	0	0	0	0	0	0
34	0	0	0	0	0	0	0	0	0	0	0	0	0	0	1	0	0	0	0	0	0	0	0	0	0	0	0	1	1	1	1	1	1	0	1	0	0	0	0	0	0	0	0	0
35	0	0	0	0	0	0	0	0	0	0	0	0	0	0	1	0	0	0	0	0	0	0	0	0	0	0	0	1	1	1	1	1	1	1	0	0	0	0	0	0	0	0	0	0
36	1	0	0	0	0	0	0	0	0	1	0	0	0	0	0	0	0	0	0	0	0	0	0	0	0	0	0	0	0	0	0	0	0	0	0	0	0	0	0	0	0	0	0	1
37	1	0	0	0	0	0	0	0	0	0	0	0	0	0	0	0	0	0	0	0	0	0	0	0	0	0	0	0	0	0	0	0	0	0	0	0	0	0	0	0	0	0	0	0
38	1	0	0	0	1	0	0	0	0	0	0	0	0	0	0	0	0	0	0	0	0	0	0	0	0	0	0	0	0	0	0	0	0	0	0	0	0	0	0	0	0	0	0	0
39	1	0	0	0	0	0	0	0	0	0	0	0	0	0	0	1	0	0	0	0	0	0	0	0	0	0	0	0	0	0	0	0	0	0	0	0	0	0	0	0	0	0	0	0
40	0	0	0	0	0	0	0	0	0	0	1	0	0	0	0	0	0	0	0	0	0	0	0	0	0	0	0	0	0	0	0	0	0	0	0	0	0	0	0	0	0	1	0	0
41	1	0	0	0	0	0	0	0	0	0	0	0	0	0	0	0	0	0	0	0	0	0	0	0	0	0	0	0	0	0	0	0	0	0	0	0	0	0	0	0	0	0	0	0
42	0	0	0	0	0	0	0	0	0	0	1	0	0	0	0	0	0	0	0	0	0	0	0	0	0	0	0	0	0	0	0	0	0	0	0	0	0	0	0	1	0	0	0	0
43	1	0	0	0	1	1	0	0	0	0	0	0	0	0	0	0	0	0	0	0	0	0	0	0	0	0	0	0	0	0	0	0	0	0	0	0	0	0	0	0	0	0	0	0
44	1	0	0	0	0	1	0	1	0	1	0	0	0	0	1	0	0	0	0	0	0	1	0	1	0	0	0	0	0	0	0	0	0	0	0	1	0	0	0	0	0	0	0	0

The Togo Network

	1	2	3	4	5	6	7	8	9	10	11	12	13	14	15	16	17	18	19	20	21	22	23	24	25	26	27	28	29	30	31	32	33
1	0	1	1	0	0	1	0	0	1	1	1	1	1	1	0	0	1	1	1	1	0	0	0	1	1	0	0	0	1	0	1	1	1
2	1	0	1	0	0	0	1	1	0	0	0	0	0	0	0	0	0	0	0	0	0	0	0	1	1	1	1	1	0	0	0	0	0
3	1	1	0	1	1	1	0	0	0	0	0	0	0	0	0	0	0	0	0	0	0	0	0	0	0	0	0	0	0	0	0	0	0
4	0	0	1	0	0	1	0	0	0	0	0	0	0	0	0	0	0	0	0	0	0	0	0	0	0	0	0	0	0	0	0	0	0
5	0	0	1	0	0	0	1	0	0	0	0	0	0	0	0	0	0	0	0	0	0	0	0	0	0	0	0	0	0	0	0	0	0
6	1	0	1	1	0	0	0	0	0	0	0	0	1	0	0	0	0	0	0	0	0	0	0	0	0	0	0	0	0	0	0	1	0
7	0	1	0	0	1	0	0	1	0	0	0	0	0	0	0	0	0	0	0	0	0	0	0	0	0	0	0	0	0	0	0	0	0
8	0	1	0	0	0	0	1	0	0	0	0	0	0	0	0	0	0	0	0	0	0	0	0	0	0	0	0	0	0	0	0	0	0
9	1	0	0	0	0	0	0	0	0	0	0	0	0	0	0	0	0	0	0	0	0	0	0	0	0	0	0	0	1	1	0	0	0
10	1	0	0	0	0	0	0	0	0	0	0	0	0	0	0	0	0	0	0	0	0	0	0	0	0	0	0	0	0	0	0	0	0
11	1	0	0	0	0	0	0	0	0	0	0	0	0	0	0	0	0	0	0	0	0	0	0	0	0	0	0	0	0	0	0	0	0
12	1	0	0	0	0	0	0	0	0	0	0	0	0	0	0	0	0	0	0	0	0	0	0	0	0	0	0	0	0	0	0	0	0
13	1	0	0	0	0	1	0	0	0	0	0	0	0	0	0	0	0	0	0	0	0	0	0	0	0	0	0	0	0	0	0	0	0
14	1	0	0	0	0	0	0	0	0	0	0	0	0	0	0	0	0	0	0	0	0	0	0	0	0	0	0	0	1	0	0	0	0
15	0	0	0	0	0	0	0	0	0	0	0	0	0	0	0	1	0	0	0	0	0	0	0	0	0	0	0	0	1	0	0	0	0
16	0	0	0	0	0	0	0	0	0	0	0	0	0	0	1	0	0	0	0	0	0	0	0	0	0	0	0	0	1	0	0	0	0
17	1	0	0	0	0	0	0	0	0	0	0	0	0	0	0	0	0	1	0	0	0	0	0	0	0	0	0	0	0	0	0	0	0
18	1	0	0	0	0	0	0	0	0	0	0	0	0	0	0	0	1	0	0	0	0	0	0	0	0	0	0	0	0	0	0	0	0
19	1	0	0	0	0	0	0	0	0	0	0	0	0	0	0	0	0	0	0	0	0	0	0	0	0	0	0	0	0	0	0	0	0
20	1	0	0	0	0	0	0	0	0	0	0	0	0	0	0	0	0	0	0	0	0	0	0	0	0	0	0	0	0	0	0	0	0
21	0	0	0	0	0	0	0	0	0	0	0	0	0	0	0	0	0	0	0	0	0	0	0	0	0	0	0	0	1	0	0	0	0
22	0	0	0	0	0	0	0	0	0	0	0	0	0	0	0	0	0	0	0	0	0	0	0	0	0	0	0	0	1	0	0	0	0
23	0	0	0	0	0	0	0	0	0	0	0	0	0	0	0	0	0	0	0	0	0	0	0	0	0	0	0	0	1	0	0	0	0
24	1	1	0	0	0	0	0	0	0	0	0	0	0	0	0	0	0	0	0	0	0	0	0	0	0	0	0	0	0	0	0	0	0
25	1	1	0	0	0	0	0	0	0	0	0	0	0	0	0	0	0	0	0	0	0	0	0	0	0	0	0	0	0	0	0	0	0
26	0	1	0	0	0	0	0	0	0	0	0	0	0	0	0	0	0	0	0	0	0	0	0	0	0	0	1	0	0	0	0	0	0
27	0	1	0	0	0	0	0	0	0	0	0	0	0	0	0	0	0	0	0	0	0	0	0	0	0	1	0	0	0	0	0	0	0
28	0	1	0	0	0	0	0	0	0	0	0	0	0	0	0	0	0	0	0	0	0	0	0	0	0	0	0	0	0	0	0	0	0
29	1	0	0	0	0	0	0	0	1	0	0	0	0	1	1	1	0	0	0	0	1	1	1	0	0	0	0	0	0	0	1	0	0
30	0	0	0	0	0	0	0	0	1	0	0	0	0	0	0	0	0	0	0	0	0	0	0	0	0	0	0	0	0	0	0	0	0
31	1	0	0	0	0	0	0	0	0	0	0	0	0	0	0	0	0	0	0	0	0	0	0	0	0	0	0	0	1	0	0	0	0
32	1	0	0	0	0	1	0	0	0	0	0	0	0	0	0	0	0	0	0	0	0	0	0	0	0	0	0	0	0	0	0	0	0
33	1	0	0	0	0	0	0	0	0	0	0	0	0	0	0	0	0	0	0	0	0	0	0	0	0	0	0	0	0	0	0	0	0

References

Abele, Gruppo (2003). *Synthetic Drugs Trafficking in Three European Cities: Major Trends and the Involvement of Organised Crime, Final Report*. Turin: Gipiangrafica.

Adler, Patricia A. (1985/1993). *Wheeling and Dealing: Ethnography of an Upper-Level Drug Dealing and Smuggling Community*, 2nd Edition. New York: Columbia University Press.

Agnew, Robert (2001). Building on the Foundation of General Strain Theory: Specifying the Types of Strain Most Likely to Lead to Crime and Delinquency, *Journal of Research in Crime and Delinquency* 38, 319–361.

Agnew, Robert (1997). Stability and Change in Crime Over the Life Course: A Strain Theory Explanation. In T.P. Thornberry (Ed.) *Developmental Theories of Crime and Delinquency* (pp. 101–132). New Brunswick: Transaction.

Agnew, Robert (1992). Foundation for a General Strain Theory of Crime and Delinquency, *Criminology* 30, 47–87.

Agnew, Robert (1984). Autonomy and Delinquency, *Sociological Perspectives* 27, 219–240.

Alain, Marc (2003). Les bandes de motards au Québec: la distinction entre crime organisé et criminels organisés. In M. Leblanc, M. Ouimet, and D. Szabo (Eds.) *Traité de criminologie empirique*, 3rd Edition (pp. 135–160). Montreal: Presses de l'Université de Montréal.

Alain, Marc (1999). *Contrabande et marché noir: des taxes en fumeé*. Montreal: Méridien.

Albert, Réka, Hawoong Jeong, and Albert-László Barabási (2000). Error and Attack Tolerance of Complex Networks, *Nature* 406, 378–382.

Archbold, Carol A. and Michael Meyer (1999). Anatomy of a Gang Suppression Unit: The Social Constructional of an Organizational Response to Gang Problems, *Police Quarterly* 2, 201–224.

Arlacchi, Pino (1983). *Mafia Business: The Mafia Ethic and the Spirit of Capitalism*. London: Verso.

Axelrod, Robert (1983). *The Evolution of Cooperation*. New York: Basic Books.

Baker, Wayne E. and Robert R. Faulkner (2004). Social Networks and Loss of Capital, *Social Networks* 26, 91–111.

Baker, Wayne E. and Robert R. Faulkner (2003). Diffusion of Fraud: Intermediate Economic Crime and Investor Dynamics, *Criminology* 41, 1173–1206.

Baker, Wayne E. and Robert R. Faulkner (1993). The Social Organization of Conspiracy: Illegal Networks in the Heavy Electrical Equipment Industry, *American Sociological Review* 58, 837–860.

Barger, Ralph (2000). *Hell's Angel*. New York: Harper Collins.
Bavelas, Alex (1948). A Mathematical Model for Group Structure, *Applied Anthropology* 7, 16–30.
Benkler, Yochai (2006). *The Wealth of Networks: How Social Production Transforms Markets and Freedom*. New Haven: Yale University Press.
Best, Joel and David F. Luckenbill (1982). *Organizing Deviance*. Englewood: Prentice-Hall.
Block, Alan (1991). *Perspectives on Organizing Crime: Essays in Opposition*. Dordrecht: Kluwer.
Block, Alan (1979). The Snowman Cometh: Coke in Progressive New York, *Criminology* 17, 75–99.
Block, Alan and William J. Chambliss (1981). *Organizing Crime*. New York: Elsevier.
Boissevain, Jeremy (1974). *Friends of Friends: Networks, Manipulators, and Coalitions*. Oxford: Basil Blackwell.
Bonacich, Phillip B. (1972). Factoring and Weighing Approaches to Status Scores and Clique Identification, *Journal of Mathematical Sociology* 2, 113–120.
Booth, Charles (1902). *Life and Labour of the People in London*. New York: Macmillan.
Borgatti, Stephen P. (2006). Identifying Sets of Key Players in a Social Network, *Computational & Mathematical Organization Theory* 12, 21–34.
Borgatti, Stephen P., Kathleen M. Carley, and David Krackhardt (2006). On the Robustness of Centrality Measures Under Conditions of Imperfect Data, *Social Networks* 28, 124–136.
Borgatti, Stephen P., Martin G. Everett, and Linton C. Freeman (2002). *Ucinet 6 for Windows*. Harvard: Analytic Technologies.
Bouchard, Martin and Pierre Tremblay (2005). Risks of Arrests Across Drug Markets: A Capture-Recapture Analysis of "Hidden" Dealer and User Populations, *Journal of Drug Issues* 34, 733–754.
Bovenkerk, Frank and Bashir Abou Chakra (2007). Terrorism and Organised Crime. In L. Holmes (Ed.) *Terrorism, Organised Crime, and Corruption* (pp. 29–41). Cheltenham: Edward Elgar.
Brantingham, Patricia L. and Paul J. Brantingham (1981). *Environmental Criminology*. Beverly Hills: Sage.
Brantingham, Paul J. and Patricia L. Brantingham (1984). *Patterns in Crime*. New York: Macmillan.
Bruinsma, Gerben and Wim Bernasco (2004). Criminal Groups and Transnational Illegal Markets, *Crime, Law, and Social Change* 41, 79–94.
Burt, Ronald S. (2005). *Brokerage and Closure: An Introduction to Social Capital*. Oxford: Oxford University Press.
Burt, Ronald S. (1992). *Structural Holes: The Social Structure of Competition*. Cambridge: Harvard University Press.
Burt, Ronald S., Joseph E. Jannotta, and James T. Mahoney (1998). Personality Correlates of Structural Holes, *Social Networks* 20, 63–87.
Calster, Patrick van (2006). Re-visiting *Mr. Nice*: On Organized Crime as Conversational Interaction, *Crime, Law, and Social Change* 45, 337–359.
Carley, Kathleen M., Ju-Sung Lee, and David Krackhardt (2001). Destabilizing Networks, *Connections* 24, 79–92.
Castells, Manuel (1998/2000). *End of Millennium*, 2nd Edition. Oxford: Blackwell.
Castells, Mauel (1996). *The Rise of the Network Society*. Oxford: Blackwell.
Chambliss, William J. (1978). *On the Take: From Petty Crooks to Presidents*. Bloomington: Indiana University Press.

Chesney, Kellow (1970). *The Victorian Underworld.* London: Maurice Temple Smith.

Chevrier, Emmanuelle (2004). The French Government's Will to Fight Organized Crime and Clean Up the Legal Professions, *Crime, Law, and Social Change* 42, 189–200.

Clarke, Ronald V. and Derek B. Cornish (1985). Modeling Offenders' Decisions: A Framework for Research and Policy, *Crime and Justice* 6, 147–185.

Cloward, Richard A. and Lloyd E. Ohlin (1960). *Delinquency and Opportunity.* Glencoe: Free Press.

Cohen, Albert K. (1955). *Delinquent Boys: The Culture of the Gang.* Glencoe: Free Press.

Coles, Nigel (2001). It's Not *What* You Know—It's *Who* You Know That Counts: Analysing Serious Crime Groups as Social Networks, *British Journal of Criminology* 41, 580–594.

Collins, Randall (2008). *Violence: A Micro-Sociological Theory.* Princeton: Princeton University Press.

Commission de police de Québec (1980). *Rapport d'enquête sur les activités des groupes de motards de Havre-Saint-Pierre, Sept-Iles, Mont-Joli, Saint-Gédéon, Sherbrooke et Asbestos.* Enquête sur le crime organisé, Éditeur officiel du Québec.

Cook, Karen S., Richard M. Emerson, Mary R. Gillmore, and Toshio Yamagishi (1983). The Distribution of Power in Exchange Networks: Theory and Experimental Results, *American Journal of Sociology* 89, 275–305.

Cornish, Derek B. (1994). The Procedural Analysis of Offending and Its Relevance for Situational Prevention. In R. Clarke's (Ed.) *Crime Prevention Studies* (pp. 151–196). Monsey: Criminal Justice Press.

Coughlin, Brenda C. and Sudhir A. Venkatesh (2003). The Urban Street Gang After 1970, *Annual Review of Sociology* 29, 41–64.

Cressey, Donald R. (1969). *Theft of the Nation: The Structure and Operations of Organized Crime in America.* New York: Harper and Row.

Cross, Rob and Andrew Parker (2004). *The Hidden Power of Social Networks.* Boston: Harvard Business School Press.

Decker, Scott H., Timothy S. Bynum, and Deborah L. Weisel (1998). A Tale of Two Cities: Gangs as Organized Crime Groups, *Justice Quarterly* 15, 395–425.

Desroches, Frederick J. (2005). *The Crime That Pays: Drug Trafficking and Organized Crime in Canada.* Toronto: Canadian Scholars Press.

Di Nicola, Andrea and Paola Zoffi (2004). Italian Lawyers and Criminal Clients: Risks and Countermeasures, *Crime, Law, and Social Change* 42, 201–225.

Dorn, Nicholas (2000). Performance Management, Indicators and Drug Enforcement: In the Crossfire or at the Crossroads. In M. Natarajan and M. Hough's (Eds.) *Crime Prevention Studies* (pp. 299–318). Monsey: Criminal Justice Press.

Dorn, Nicholas, Lutz Oette, and Simone White (1998). Drug Importation and the Bifurcation of Risk: Capitalization, Cut Outs, and Organized Crime, *British Journal of Criminology* 38, 537–560.

Dorn, Nicholas, Karim Murji, and Nigel South (1992). *Traffickers: Drug Markets and Law Enforcement.* London: Routledge.

Dupont, Benoît. 2006. *Les polices au Québec.* Paris: Presses Universitaires de France.

Duyne, Petrus C. van (2004). The Creation of a Threat Image: Media, Policy Making and Organised Crime. In P. Van Duyne, M. Jager, K. von Lampe, and J. Newell (Eds.) *Threats and Phantoms of Organised Crime, Corruption and Terrorism* (pp. 21–50). Nijmegen: Wolf Legal Publishers.

Duyne, Petrus C. van and Michael Levi (2005). *Drugs and Money: Managing the Drug Trade and Crime-Money in Europe.* London: Routledge.

Ebel, Holger, Jörn Davidsen, and Stefan Bornholdt (2003). Dynamics of Social Networks, *Complexity* 8, 24–27.

Eck, John E. and Jeffrey S. Gersh (2000). Drug Trafficking as a Cottage Industry. In M. Natarajan and M. Hough's (Eds.) *Crime Prevention Studies* (pp. 241–271). Monsey: Criminal Justice Press.

Ekblom, Paul (1999). Can We Make Crime Prevention Adaptive by Learning From Other Evolutionary Struggles, *Studies on Crime and Crime Prevention* 8, 27–51.

Ekblom, Paul and Nick Tilley (2000). Going Equipped: Criminology, Situational Crime Prevention and the Resourceful Offender, *British Journal of Criminology* 40, 376–398.

Erickson, Bonnie (1981). Secret Societies and Social Structure, *Social Forces* 60, 188–210.

Fagan, Jeffrey (1990). Social Processes of Delinquency and Drug Use Among Urban Gangs. In C.R. Huff (Ed.) *Gangs in America* (pp. 183–219). Sage: Newbury Park.

Fagan, Jeffrey (1989). The Social Organization of Drug Use and Drug Dealing Among Urban Gangs, *Criminology* 27, 633–665.

Farley, Jonathan David (2003). Breaking Al Qaeda Cells: A Mathematical Analysis of Counterterrorism Operations (A Guide for Risk Assessment and Decision Making), *Studies in Conflict and Terrorism* 26, 399–411.

Farrell, Graham, Kashfia Mansur, and Melissa Tullis (1996). Cocaine and Heroin in Europe, 1983–1993: A Cross-National Comparison of Trafficking and Prices, *British Journal of Criminology* 36, 255–281.

Felson, Marcus (2006). *Crime and Nature*. Thousand Oaks: Sage.

Felson, Marcus (2003). The Process of Co-offending. In M.J. Smith and D.B. Cornish (Eds.) *Crime Prevention Studies* (pp. 149–167). Monsey: Criminal Justice Press.

Finckenauer, James O. and Elin J. Waring (1998). *Russian Mafia in America: Immigration, Culture, and Crime*. Boston: Northeastern University Press.

Fischer, Claude S. (1977). Network Analysis and Urban Studies. In C.S. Fischer, R.M. Jackson, C.A. Stueve, K. Gerson, L.M. Jones, and M. Baldassare (Eds.) *Networks and Places: Social Relations in the Urban Setting* (pp. 19–37). New York: Free Press.

Fleisher, Mark S. (1995). *Beggars and Thieves: Lives of Urban Street Criminals.* Madison: University of Wisconsin Press.

Frank, Ove and Peter J. Carrington (2007). Estimation of Offending and Co-offending Using Available Data with Model Support, *Journal of Mathematical Sociology* 31, 1–46.

Freeman, Linton C. (2004). *The Development of Social Network Analysis: A Study in the Sociology of Science*. Vancouver: Empirical Press.

Freeman, Linton C. (1979). Centrality in Social Networks: Conceptual Clarification, *Social Networks* 1, 215–239.

Freeman, Linton C. (1977). A Set of Measures of Centrality Based on Betweenness, *Sociometry* 40, 35–41.

Freeman, Linton C., Stephen P. Borgatti, and Douglas R. White (1991). Centrality in Valued Graphs: A Measure of Betweenness Based on Network Flow, *Social Networks* 13, 141–154.

Gambetta, Diego (1993). *The Sicilian Mafia: The Business of Private Protection.* Cambridge: Harvard University Press.

Gambetta, Diego and Peter Reuter (1995). Conspiracy Among the Many: The Mafia in Legitimate Industries. In G. Fiorentini and S. Peltzman (Eds.) *The Economics of Organized Crime* (pp. 116–136). Cambridge: Cambridge University Press.

Gottfredson, Michael R. and Travis Hirschi (1990). *A General Theory of Crime*. Stanford: Stanford University Press.

Gould, Roger V. (2003). *Collision of Wills: How Ambiguity about Social Rank Breeds Conflict*. Chicago: University of Chicago Press.

Gould, Roger V. and Roberto M. Fernandez (1989). Structures of Mediation: A Formal Approach to Brokerage in Transaction Networks. In C. Clogg's (Ed.) *Sociological Methodology* (pp. 89–126). Oxford: Basil Blackwell.

Hagan, John (1997). Defiance and Despair: Subcultural and Structural Linkages between Delinquency and Despair in the Life Course, *Social Forces* 76, 119–134.

Haller, Mark H. (1990). Illegal Enterprise: A Theoretical and Historical Interpretation, *Criminology* 28, 207–235.

Hayek, Friedrich A. (1973). *Law, Legislation, and Liberty, Volume 1: Rules and Order*. Chicago: University of Chicago Press.

Haynie, Dana L. (2001). Delinquent Peers Revisited: Does Network Structure Matter? *American Journal of Sociology* 106, 1013–1057.

Heckathorn, Douglas D. and Joan Jeffri (2001). Finding the Beat: Using Respondent-Driven Sampling to Study Jazz Musicians, *Poetics* 28, 307–329.

Hill, Peter B.E. (2003). *The Japanese Mafia: Yakuza, Law, and the State*. Oxford: Oxford University Press.

Hobbs, Dick (2001). The Firm: Organizational Logic and Criminal Culture on a Shifting Terrain, *British Journal of Criminology* 41, 549–560.

Hobsbawm, Eric J. (1969). *Bandits*. London: Weidenfeld and Nicholson.

Hobsbawm, Eric J. (1959). *Primitive Rebels: Studies in Archaic Forms of Social Movement in the 19th and 20th Centuries*. New York: W.W. Norton and Company.

Hoffer, Lee D. (2006). *Junkie Business: The Evolution and Operation of a Heroin Dealing Network*. Belmont: Thomson/Wadsworth.

Hubbell, Charles H. (1965). An Input-Output Approach to Clique Identification, *Sociometry* 28, 377–399.

Huff, C. Ronald (1990). Denial, Overraction, and Misidentification: A Postscript of Public Policy. In C.R. Huff (Ed.) *Gangs in America* (pp. 310–317). Newbury Park: Sage.

Ianni, Francis J. (1972). *A Family Business*. New York: Russell Sage Foundation.

Jackson, Pat and Cary Rudman (1993). Moral Panic and the Response to Gangs in California. In S. Cummings and D.J. Monti (Eds.) *Gangs: The Origins and Impact of Contemporary Youth Gangs in the United States* (pp. 257–275). Albany: State University of New York Press.

Jacobs, Bruce A. (1999). *Dealing Crack: The Social World of Streetcorner Selling*. Boston: Northeastern University Press.

Jacobs, James and Ellen Peters (2003). Labor Racketeering: The Mafia and the Unions, *Crime and Justice: A Review of Research* 30, 229–281.

Jacobs, James, Coleen Friel, and Robert Radick (1999). *Gotham Unbound*. New York: New York University Press.

Jennings, Helen H. (1943). *Leadership and Isolation: A Study of Personality in Inter-Personal Relations*. New York: Longmans, Green, and Company.

Kalish, Yuval and Garry Robins (2006). Psychological Predispositions and Network Structure: The Relationship Between Individual Predispositions, Structural Holes, and Network Closure, *Social Networks* 28, 56–84.

Katz, Charles (2003). Issues in the Production and Dissemination of Gang Statistics: An Ethnographic Study of a Large Midwestern Police Gang Unit, *Crime and Delinquency* 49, 485–516.

Katz, Charles and Vincent J. Webb (2006). *Policing Gangs in America.* New York: Cambridge University Press.

Katz, Charles, Vincent J. Webb, and David R. Schaefer (2000). The Validity of Police Gang Intelligence Lists: Examining Differences in Delinquency Between Documented Gang Members and Nondocumented Delinquent Youth, *Police Quarterly* 3, 413–437.

Kennedy, David M., Anthony A. Braga, and Anne M. Piehl (1997). The (Un)Known Universe: Mapping Gangs and Gang Violence in Boston. In David L. Weisburd and J. Thomas McEwen (Eds.) *Crime Mapping and Crime Prevention* (pp. 219–262). New York: Criminal Justice Press.

Kick, Edward L., and Byron L. Davis (2001). World-System Structure and Change: An Analysis of Global networks and Economic Growth Across Two Time Periods, *American Behavioral Scientist* 44, 1561–1578.

Kleemans, Edward R. (2007). Organized Crime, Transit Crime, and Racketeering, *Crime and Justice: A Review of Research* 35, 163–215.

Kleemans, Edward R. and Christianne J. dePoot (2008). Criminal Careers in Organized Crime and Social Opportunity Structure, *European Journal of Criminology* 5, 69–98.

Kleemans, Edward R. and Henk G. Van de Bunt (2003). The Social Organisation of Human Trafficking. In D. Siegel, H.G. van de Bunt, and D. Zaitch (Eds.) *Global Organized Crime: Trends and Developments* (pp. 97–104). Dordrecht: Kluwer.

Kleemans, Edward R. and Henk G. Van de (1999). The Social Embeddedness of Organized Crime. *Transnational Organized Crime* 5, 19–36.

Klein, Malcolm W. (1971). *Street Gangs and Street Workers.* Englewood Cliffs: Prentice-Hall.

Klein, Malcolm W. and Cheryl L. Maxson (2006). *Street Gang Patterns and Policies.* Oxford: Oxford University Press.

Klein, Malcolm W. and Lois Y. Crawford (1967). Groups, Gangs, and Cohesiveness, *Journal of Research in Crime and Delinquency* 4, 63–75.

Klerks, Peter (2001). The Network Paradigm Applied to Criminal Organizations, *Connections* 24, 53–65.

Krebs, Valdis (2001). Mapping Networks of Terrorist Cells, *Connections* 24, 43–52.

Krebs, Valdis (2002). Uncloaking Terrorist Networks, *First Monday* 7.

Lacoste, Julie and Pierre Tremblay (2003). Crime and Innovation: A Script Analysis of Patterns in Check Forgery. In M.J. Smith and D.B. Cornish (Eds.) *Crime Prevention Studies* (pp. 169–196). Monsey: Criminal Justice Press.

Lampe, Klaus von (2003). Criminally Exploitable Ties: A Network Approach to Organized Crime. In E.C. Viano, J. Magallanes, and L. Bidel (Eds.) *Transnational Organized Crime: Myth, Power and Profit* (pp. 9–22). Durham: Carolina Academic Press.

Lankhorst, Francien and Hans Nelen (2004). Professional Services and Organised Crime in the Netherlands, *Crime, Law, and Social Change* 42, 163–188.

Laumann, Edward O., Peter V. Marsden, and David Prensky (1983). The Boundary Specification Problem in Social Networks. In R.S. Burt, M.J. Minor (Eds.) *Applied Network Analysis* (pp. 18–34). London: Sage.

Levi, Michael, Hans Nelen, and Francien Lankhorst (2004). Lawyers as Crime Facilitators in Europe: An Introduction and Overview, *Crime, Law, and Social Change* 42, 117–121.

Levitt, Steven D. and Sudhir A. Ventatesh (2000). An Economic Analysis of a Drug-Selling Gang's Finances, *The Quarterly Journal of Economics* 115, 755–789.

Liddick, Don (1999). The Enterprise "Model" of Organized Crime: Assessing Theoretical Propositions, *Justice Quarterly* 16, 403–430.

Lyman, Michael and Gary W. Potter (2000). *Organized Crime*, 2nd Edition. New York: Prentice-Hall.

MacRae, Duncan Jr. (1960). Direct Factor Analysis of Sociometric Data, *Sociometry* 23, 360–371.

Marks, Howard (1997). *Mr. Nice: An Autobiography*. London: Minerva.

Mars, Gerald (1983). *Cheats at Work: An Anthropology of Workplace Crime*. London: Allen and Unwin.

Marsden, Peter V. (1982). Brokerage Behavior in Restricted Exchange Networks. In P.V. Marsden and N. Lin (Eds.) *Social Structure and Network Analysis* (pp. 201–218). London: Sage.

Mayhew, Henry (1861/1968) *London Labour and the London Poor*. New York: A.M. Kelley.

McAndrew, Duncan (1999). The Structural Analysis of Criminal Networks. In D. Canter and L.J. Alison (Eds.) *The Social Psychology of Crime: Groups, Teams, and Networks* (pp. 52–92). Aldershot: Ashgate.

McCorkle, Richard C. and Terance D. Miethe (1998). The Political and Organizational Response to Gangs: An Examination of a "Moral Panic" in Nevada, *Justice Quarterly* 15, 41–64.

McGloin, Jean Marie (2007). The Organizational Structure of Street Gangs in Newark, New Jersey: A Network Analysis Methodology, *Journal of Gang Research* 15, 1–34.

McGloin, Jean Marie (2005). Police and Intervention Considerations of a Network Analysis of Street Gangs, *Criminology and Public Policy* 4, 607–635.

McGloin, Jean Marie, Christopher J. Sullivan, Alex R. Piquero, and Sarah Bacon (2008). Investigating the Stability of Co-offending and Co-offenders Among a Sample of Youthful Offenders, *Criminology* 46, 155–188.

McIllwain, Jeffrey S. (2004). *Organizing Crime in Chinatown: Race and Racketeering in New York City, 1890–1910*. Jefferson: McFarland and Company.

McIllwain, Jeffrey S. (1999). Organized Crime: A Social Network Approach, *Crime, Law and Social Change* 32, 301–324.

Middleton, David H. and Michael Levi (2004). The Role of Solicitors in Facilitating 'Organized Crime': Situational Crime Opportunities and Their Regulation, *Crime, Law, and Social Change* 42, 123–161.

Milgram, Stanley (1967). The Small World Problem, *Psychology Today* 1, 61–67.

Milhaupt, Curtis J. and Mark D. West (2000). The Dark Side of Private Ordering: An Institutional and Empirical Analysis of Organized Crime, *University of Chicago Law Review* 67, 41–98.

Moody, James (2001). Peer Influence Groups: Identifying Dense Clusters in Large Networks, *Social Networks* 23, 261–283.

Moreno, Jacob L. (1957). *The First Book on Group Psychotherapy*. New York: Psychodrama and Group Psychotherapy Monographs 1.

Moreno, Jacob L. (1934/1953). *Who Shall Survive: Foundations of Sociometry, Group Psychotherapy and Sociodrama*. Beacon: Beacon House.

Moreno, Jacob L. (1932). *Application of the Group Method to Classification*. New York: National Committee on Prisons and Prison Labor.

Morselli, Carlo (2005). *Contacts, Opportunities, and Criminal Enterprise*. Toronto: University of Toronto Press.

Morselli, Carlo, Dave Tanguay, and Anne-Marie Labalette (2008). Criminal Conflicts and Collective Violence: Biker-Related Account Settlements in Quebec, 1994–2001. In D. Siegel and H. Nelen (Eds.) *Organized Crime: Culture, Markets, and Policies* (pp. 145–164). New York: Springer.

Morselli, Carlo, Pierre Tremblay, and Bill McCarthy (2006). Mentors and Criminal Achievement, *Criminology* 44, 17–44.

Morselli, Carlo and Pierre Tremblay (2004). Criminal Achievement, Offender Networks, and the Benefits of Low Self Control, *Criminology* 42, 773–804.

Nakamura, Kiminori, George Tita, and David Krackhardt (2006). Testing Balance Theory: Examination of Gang Networks of Rivals and Allies. Paper Presented at the Annual Meeting of the *American Society of Criminology*, Los Angeles.

Natarajan, Mangai (2006). Understanding the Structure of a Large Heroin Distribution Network: A Quantitative Analysis of Qualitative Data. *Journal of Quantitative Criminology* 22, 171–192.

Natarajan, Mangai (2000). Understanding the Structure of a Drug Trafficking Organization: A Conversational Analysis. In M. Natarajan and M. Hough (Eds.) *Crime Prevention Studies* (pp. 273–298). Monsey: Criminal Justice Press.

Naylor, Robin Thomas (2006). *Satanic Purses: Money, Myth, and Misinformation in the War on Terror*. Montreal: McGill-Queen's University Press.

Naylor, Robin Thomas (2002). *Wages of Crime: Black Markets, Illegal Finance, and the Underground Economy*. Ithaca: Cornell University Press.

Naylor, R. Thomas (1997). Mafias, Myths, and Markets: On the Theory and Practice of Organized Crime, *Transnational Organized Crime* 3, 1–45.

Papachristos, Andrew V. (2007). Murder by Structure: Dominance Relations and the Social Structure of Gang Homicide in Chicago. *Social Science Research Network*, Working Paper.

Papachristos, Andrew V. (2006). Social Network Analysis and Gang Research. In J.F. Short and L.A. Hughes (Eds.) *Studying Youth Gangs* (pp. 99–116). Walnut Creek: Alta Mira.

Pearson, Geoffrey and Dick Hobbs (2001). *Middle Market Drug Distribution*. London: Home Office.

Peterson, Marilyn (1994). *Applications in Criminal Analysis: A Sourcebook*. Westport: Greenwood Press.

Piquero, Alex R., David P. Farrington, and Alfred Blumstein (2007). *Key Issues in Criminal Careers: New Analyses of the Cambridge Study in Delinquent Development*. Cambridge: Cambridge University Press.

Piquero, Alex R., David P. Farrington, and Alfred Blumstein (2003). The Criminal Career Paradigm: Background and Recent Developments, *Crime and Justice: A Review of Research* 30, 137–183.

Podolny, Joel M. and Karen L. Page (1998). Network Forms of Organization, *Annual Review of Sociology* 24, 57–76.

Potter, Gary W. (1994). *Criminal Organizations: Vice, Racketeering, and Politics in an American City*. Prospect Heights: Waveland Press.

Powell, Walter W. (1990). Neither Market Nor Hierarchy: Network Forms of Organization. In B. Staw (Ed.) *Research in Organizational Behavior* (pp. 295–293). Greenwich: JAI Press.

Raab, Jörg and H. Brinton Milward (2003). Dark Networks as Problems. *Journal of Administration Research and Theory* 13, 413–439.

Reiss, Albert J. (1986). Co-Offender Influences on Criminal Careers. In A. Blumstein, J. Cohen, J. Roth, and C. Visher (Eds.) *Criminal Careers and 'Career Criminals'* (pp. 121–160). Washington D.C.: National Academic Press.

Reuter, Peter (1983). *Disorganized Crime: The Economics of the Invisible Hand*. Cambridge: MIT Press.

Reuter, Peter and John Haaga (1989). *The Organization of High-Level Drug Markets: An Exploratory Study*. Santa Monica: Rand.

Reuter, Peter and Mark A. R. Kleiman (1986). Risks and Prices: An Economic Analysis of Drug Enforcement, *Crime and Justice: An Annual Review of Research* 7, 289–340.

Rossem, R. van (1996). The World System Paradigm as General Theory of Development: A Cross-National Test, *American Sociological Review* 61, 508–527.

Ruggiero, Vincenzo (1996). *Organized Crime and Corporate Crime in Europe*. Darthmouth: Aldershot.

Salganik, Matthew J. and Douglas D. Heckathorn (2004). Sampling and Estimation in Hidden Populations Using Respondent-Driven Sampling, *Sociological Methodology* 34, 193–239.

Sanders, William B. (1994). *Gangbangs and Drive-bys: Grounded Culture and Juvenile Gang Violence*. New York: Aldine de Gruyter.

Sanger, Daniel (2005). *Hell's Witness*. Toronto: Viking Canada.

Sarnecki, Jerzy (2001). *Delinquent Networks: Youth Co-offending in Stockholm*. Cambridge: Cambridge University Press.

Sarnecki, Jerzy (1986). *Delinquent Networks*. Stockholm: National Council for Crime Prevention.

Schmid, Alex P. (1996). The Links Between Transnational Organized Crime and Terrorist Crimes, *Transnational Organized Crime* 2, 40–82.

Scott, John (1991). *Social Network Analysis: A Handbook*. London: Sage.

Shaw, Clifford R. and McKay, Henry D. (1942). *Juvenile Delinquency and Urban Areas*. Chicago: The University of Chicago Press.

Sher, Julian and William Marsden (2004). *The Road to Hell: How the Bikers are Conquering Canada*. Toronto: Vintage Canada.

Sherman, Lawrence W., Patrick R. Gartin, and Michael E. Buerger (1989). Hot Spots of Predatory Crime: Routine Activities and the Criminology of Place, *Criminology* 27, 27–55.

Short, James F. and Fred L. Strodtbeck (1965). *Group Processes and Gang Delinquency*. Chicago: University of Chicago Press.

Simmel, Georg (1908/1950). The Secret and the Secret Society. In K.H. Wolff (Ed.) *The Sociology of Georg Simmel* (pp. 307–338). New York: Free Press.

Skolnick, Jerome H., Ricky Bluthenthal, and Theodore Correl (1993). Gang Organization and Migration. In S. Cummings and D.J. Monti (Eds.) *Gangs: The Origins and Impact of Contemporary Youth Gangs in the United States* (pp. 193–217). Albany: State University of New York Press.

Smith, David A. and Michael F. Timberlake (2001). World City Networks and Hierarchies, 1977–1997, *American Behavioral Scientist* 44, 1656–1678.

Smith, David A. and Douglas R. White (1992). Structure and Dynamics of the Global Economy: Network Analysis of International Trade, 1965–1980, *Social Forces* 70, 857–893.

Smith, Dwight C. (1980). Paragons, Pariahs, and Pirates: A Spectrum-Based Theory of Enterprise, *Crime and Delinquency* 26, 358–386.

Snyder, David and Edward L. Kick (1979). Structural Position in the World System and Economic Growth, 1955–1970: A Multiple Network Analysis of Transnational Interaction, *American Journal of Sociology* 84, 1096–1126.

Sparrow, Malcolm K. (1991). The Application of Network Analysis to Criminal Intelligence: An Assessment of the Prospects, *Social Networks* 13, 251–274.

Spergel, Irving A. (1995). *The Youth Gang Problem: A Community Approach*. New York: Oxford University Press.

Steffensmeier, Darrell (1986). *The Fence: In the Shadow of Two Worlds*. Totowa: Rowman and Littlefield.

Steffensmeier, Darrell and Jeffery T. Ulmer (2005). *Confessions of a Dying Thief: Understanding Criminal Careers and Illegal Enterprise*. New Brunswick: Transaction.

Sutherland, Edwin H. (1947). *Principles of Criminology*, 4th Edition. New York: Lippincott.

Suttles, Gerald D. (1968). *The Social Order of the Slum: Ethnicity and Territory in the Inner City*. Chicago: University of Chicago Press.

Sykes, Gresham M. (1958). *The Society of Captives: A Study of a Maximum Security Prison*. Princeton: Princeton University Press

Thornberry, Terence P., Marvin D. Krohn, Alan J. Lizotte, Carolyn A. Smith, and Kimberly Tobin (2003). *Gangs and Delinquency in Developmental Perspective*. New York: Cambridge University Press.

Thrasher, Frederic M. (1927). *The Gang: A Study of 1,313 Gangs in Chicago*. Chicago: Chicago University Press.

Tita, George, K. Jack Riley, Greg Ridgeway, Clifford Grammich, Allan F. Abrahamse, and Peter W. Greenwood (2003). *Reducing Gun Violence: Results from an Intervention in East Los Angeles*. Santa Monica: Rand.

Tremblay, Pierre (1993). Searching for Suitable Co-offenders. In R.V. Clarke and M. Felson (Eds.) *Routine Activity and Rational Choice: Advances in Criminological Theory* (pp. 17–36). New Brunswick: Transaction.

Tremblay, Pierre, Bernard Talon, and Doug Hurley (2001). Body Switching and Related Adaptations in the Resale of Stolen Vehicles, *British Journal of Criminology* 41, 561–579.

Tremblay, Pierre, Maurice Cusson, and Carlo Morselli (1998). Market Offenses and Limits to Growth, *Crime, Law, and Social Change* 29, 311–330.

Tremblay, Pierre, Sylvie Laisne, Gilbert Cordeau, Brian MacLean, and Angela Shewshuck (1989). Carrières criminelles collectives: évolution d'une population délinquante (les groupes motards), *Criminologie* 22, 65–94.

Varese, Federico (2006). The Structure of a Criminal Network Examined: The Russian Mafia in Rome, *Oxford Legal Studies Research Paper* 21.

Varese, Federico (2001). *The Russian Mafia: Private Protection in a New Market Economy*. Oxford: Oxford University Press.

Venkatesh, Sudhir A. and Steven D. Levitt (2000). "Are We a Family or a Business?" History and Disjuncture in the Urban American Street Gang, *Theory and Society* 29, 427–462.

Vigil, James D. (1988). *Barrio Gangs: Street Life and Identity in Southern California*. Austin: University of Texas Press.

Wallerstein, Immanuel (1974). *The Modern World-System*. New York: Academic Press.

Walsh, Marilyn E. (1977). *The Fence: A New Look at the World of Property Theft*. Westport: Greenwood Press.

Waring, Elin J. (1993). *Co-offending in White Collar Crime: A Network Approach*. Ann Arbor: University Microfilms International.

Wasserman, Stanley and Katherine Faust (1994). *Social Network Analysis: Methods and Applications*. Cambridge: Cambridge University Press.

Watts, Duncan J. (2003). *Six Degrees: The Science of a Connected Age*. New York: W.W. Norton and Company.

Watts, Duncan J. (1999). *Small Worlds: The Dynamics of Networks Between Order and Randomness*. Princeton: Princeton University Press.

Weerman, Frank M. and Wilma H. Smeenk (2005). Peer Similarity in Delinquency for Different Types of Friends: A Comparison Using Two Measurement Methods, *Criminology* 43, 499–523.

Weisel, Deborah L. (2002a). *Contemporary Gangs: An Organizational Analysis.* New York: LFB Scholarly Publishing.

Weisel, Deborah L. (2002b). The Evolution of Street Gangs: An Examination of Form and Variation. In W.L. Reed and S.H. Decker (Eds.) *Responding to Gangs: Evaluation and Research* (pp. 25–65). Washington D.C.: National Institute of Justice.

Whyte, William F. (1943/1993). *Street Corner Society: The Social Structure of an Italian Slum*. Chicago: University of Chicago Press.

Williams, Phil (2001). Transnational Criminal Networks. In J. Arquilla and D. Ronfeldt (Eds.) *Networks and Netwars* (pp. 61–97). Santa Monica: Rand.

Williams, Phil (1998). The Nature of Drug-Trafficking Networks, *Current History* 97, 154–159.

Wolf, Daniel R. (1991). *The Rebels: A Brotherhood of Outlaw Bikers*. University of Toronto Press: Toronto.

Woodiwiss, Michael (2001). *Organized Crime and American Power: A History*. Toronto: University of Toronto Press.

Xu, Jennifer, Byron Marshall, Siddhardt Kaza, and Hsinchun Chen (2004). Analyzing and Visualizing Criminal Network Dynamics: A Case Study, *Intelligence and Security Informatics, Proceedings* 3073, 359–377.

Yablonsky, Lewis (1962). *The Violent Gang*. Baltimore: Penguin.

Zaitch, Damián (2002). *Trafficking Cocaine: Colombian Drug Entrepreneurs in the Netherlands*. The Hague: Kluwer Law International.

Zatz, Marjorie S. (1987). Chicano Youth Gangs and Crime: The Creation of a Moral Panic, *Crime, Law, and Social Change* 11, 129–158.

Zhang, Sheldon (2008). *Chinese Human Smuggling Organizations: Families, Social Networks, and Cultural Imperatives*. Palo Alto: Stanford University Press.

Zhang, Sheldon and Ko-lin Chin (2002). Enter the Dragon: Inside Chinese Human Smuggling Operations, *Criminology* 40, 737–767.

Index

FSC
www.fsc.org
MIX
Papier | Fördert
gute Waldnutzung
FSC® C083411

Kolibri 360 GmbH
Ferdinand-Jühlke-Straße 7
99095 Erfurt, Deutschland
produktsicherheit@kolibri360.de